New Model of Burn Out Syndrome: Towards Early Diagnosis and Prevention

RIVER PUBLISHERS SERIES IN RESEARCH AND BUSINESS CHRONICLES: BIOTECHNOLOGY AND MEDICINE

Volume 1

Combining a deep and focused exploration of areas of basic and applied science with their fundamental business issues, the series highlights societal benefits, technical and business hurdles, and economic potentials of emerging and new technologies. In combination, the volumes relevant to a particular focus topic cluster analyses of key aspects of each of the elements of the corresponding value chain.

Aiming primarily at providing detailed snapshots of critical issues in biotechnology and medicine that are reaching a tipping point in financial investment or industrial deployment, the scope of the series encompasses various specialty areas including pharmaceutical sciences and healthcare, industrial biotechnology, and biomaterials. Areas of primary interest comprise immunology, virology, microbiology, molecular biology, stem cells, hematopoiesis, oncology, regenerative medicine, biologics, polymer science, formulation and drug delivery, renewable chemicals, manufacturing, and biorefineries.

Each volume presents comprehensive review and opinion articles covering all fundamental aspect of the focus topic. The editors/authors of each volume are experts in their respective fields and publications are peer-reviewed.

For a list of other books in this series, www.riverpublishers.com
http://riverpublishers.com/series.php?msg=Research and Business Chronicles: Biotechnology and Medicine

New Model of Burn Out Syndrome: Towards Early Diagnosis and Prevention

Editor

Drozdstoy St. Stoyanov, MD, PhD, PGCert

Full Professor of Psychiatry, Medical Psychology
and Person Centered Medicine,
Medical University,
Plovdiv, Bulgaria

River Publishers
Aalborg

Published, sold and distributed by:
River Publishers
Niels Jernes Vej 10
9220 Aalborg Ø
Denmark

ISBN: 978-87-93102-70-5 (Print)
978-87-93102-71-2 (Ebook)

Foreword

Burn-out is a psychological disengagement from life characterized by emotional exhaustion, social alienation, and reduced achievement. The phenomenon of burnout is an increasing source of personal demoralization and ineffectiveness in professional settings like medical centers. This growing problem in modern society leads to unprofessional and ineffective behaviors, such as anxiety, irritability, impatience, cynicism, and lack of empathy. This monograph is designed to begin the process of correcting the problem of burnout by understanding its causes in terms of those aspects of an individual's personality that make him or her vulnerable to the stresses of an institution's psychological climate. We need to understand people in their psychosocial context in order to be able to plan interventions that can help to prevent or relieve burn-out in the workplace.

Much work on burn-out has focused on characterizing its psychopathology and resulting exhaustion, cynicism, and inefficacy. This monograph emphasizes the need to characterize both the strengths and the weaknesses of individuals that influence their overall well-being, including physical, emotional, social, cognitive, and spiritual aspects of health (Cloninger and Zohar 2011; Eley, Cloninger et al. 2013). The authors are able to characterize the whole person by using the Temperament and Character Inventory (TCI), which distinguishes between a person's emotional drives (that is, their temperament) and their capacity for effective mental self-government (that is, their character) (Cloninger, Svrakic et al. 1993). Much prior work has shown that disorders like burn-out that are characterized by anxious and depressive symptoms are likely to be associated with hypersensitivity to aversive stimuli as measured by the TCI dimension of Harm Avoidance (i.e., being pessimistic, fearful, shy, and fatigable) (Cloninger, Zohar et al. 2010). However, our tendency to become exhausted and cynical in response to punishment and frustrative non-reward can be modulated by healthy character traits and persistence (Cloninger, Zohar et al. 2012; Eley, Cloninger et al. 2013; Eley, Wilkinson et al. 2013). In addition, the psychological climate to which the person must adapt must be taken into consideration because people are inseparable from their environment (Cloninger 2004).

These questions require us to ask what are the characteristics of a healthy and resilient person? We know that healthy character traits include mature development of Self-directedness (i.e., purposeful and resourceful), Cooperativeness (i.e., helpful and forgiving), and Self-transcendence (i.e., unselfish and virtuous), but how many of us know how to develop and maintain these characteristics under the stress of modern life? (Cloninger 2004; Cloninger 2013). We need to know more about what are the characteristics of a salutogenic and supportive institutional environment, as is carefully considered in this monograph. Surely these are among the most important questions that all of us need to ask in facing the challenges of modern life.

Fortunately the authors have begun to paint a clear picture of healthy people, salutogenic institutional climates, and what a person can do to develop greater resilience and well-being (Eley, Wilkinson et al. 2013). What we see emerging is a set of observations that indicate that personal well-being is an inseparable component of the collective well-being of institutions (Cloninger 2013). The take-home message is that each of us needs to be both generous and self-reliant, thereby working with generosity and kindness in a way that is respectful of others and their intrinsic value as human beings, thereby creating a mutually supportive and secure psychological climate for one another. If we view ourselves as separate and live defensively or apprehensively, we are vulnerable to disengagement from work and other people in a way that deprives us of the experiences that give our life its meaning and satisfaction.

C. Robert Cloninger, MD, PhD
Wallace Renard Professor of Psychiatry, Genetics, & Psychology
Washington University in St. Louis

C. Robert Cloninger is Wallace Renard Professor of Psychiatry, Professor of Psychology and Genetics, and Director of the Sansone Center for Well-Being at Washington University Medical School in St. Louis. He received his B. A. with High Honors and Special Honors in Philosophy, Psychology, and Anthropology from the University of Texas at Austin, 1966. He received his M. D. from Washington University in 1970, and Honorary Doctorates from the University of Umea in 1983 (MD in Genetics) and University of Gothenburg in 2012 (PhD in Psychology). His recent books include *Feeling Good*: *The Science of Well-Being* by Oxford University Press, *Origins of Altruism and Cooperation* by Springer, and *Personality and Psychopathology* by American Psychiatric Press. Dr. Cloninger has received the American Psychiatric Association's Adolf Meyer Award (1993) for contributions to psychobiology and Judd Marmor Award (2009) for contributions to the bio-psychosocial basis of mental health. He received lifetime achievement awards from the American Society of Addiction Medicine (2000) and the International Society of Psychiatric Genetics (2003). He received the Oskar Pfister Award in 2014 from the American Association of Professional Chaplains and the American Psychiatric Association for his contributions to dialogue between psychiatry, religion, and spirituality.

References

Cloninger, C. R. (2004). Feeling Good: The Science of Well-Being. New York, Oxford University Press.

Cloninger, C. R. (2013). "What makes people healthy, happy, and fulfilled in the face of current world challenges? " Mens Sana Monogr 11: 16–24.

Cloninger, C. R., D. M. Svrakic, et al. (1993). "A psychobiological model of temperament and character. " Arch Gen Psychiatry 50(12): 975–90.

Cloninger, C. R. and A. H. Zohar (2011). "Personality and the perception of health and happiness. " J Affect Disord 128(1–2): 24–32.

Cloninger, C. R., A. H. Zohar, et al. (2010). "Promotion of well-being in person-centered mental health care. " Focus 8(2): 165–179.

Cloninger, C. R., A. H. Zohar, et al. (2012). "The psychological costs and benefits of being highly persistent: personality profiles distinguish mood disorders from anxiety disorders. " J Affect Disord 136(3): 758–66.

Eley, D. S., C. R. Cloninger, et al. (2013). "The relationship between resilience and personality traits in doctors: implications for enhancing well-being. " PeerJ 1: e216.

Eley, D. S., D. Wilkinson, et al. (2013). "Physician heal thyself, and develop your resilience. " British Medical Journal Careers.

Contents

List of the Most Common Abbreviations

BOS – Burn Out Syndrome

C - Cooperativeness

DP – Depersonalization

EE – Emotional Exhaustion

HA – Harm Avoidance

IMPC – Inductive Measures of Psychological Climate

MBI – Maslach Burn Out Inventory

NS – Novelty Seeking

P - Persistemce

PA – Personal Accomplishement

RD – Reward Dependence

SD – Self-directedeness

ST – Self-transcendence

TCI-R – Temperament and Character Inventory Revised

List of Figures

List of Tables

1

Introduction and Background of the Studies in the Field of Burn Out Syndrome (BOS)

Maria Stoykova DD, PhD, Associate Professor
Head of the Department of Social Medicine and Public Health,
Department of Social Medicine and Public Health,
Faculty of Public Health, Medical University of Plovdiv

Stanislava Harizanova MD, PhD, Assistant Professor,
Department of Hygiene and Ecomedicine,
Faculy of Public Health, Medical University of Plovdiv

Drozdstoj Stoyanov MD, PhD Full Professor of Psychiatry,
Medical Psychology and Person Centered Medicine
Faculty of Medicine, MUP, Vice Chair Executive Committee PSIG,
Royal College of Psychiatrists, Visiting Fellow, University of Pittsburgh

Burn out draws attention to researchers from a long time. Probably the earliest written example in which "to burn out" is related to exhaustion comes from Shakespeare, who wrote in 1599 in The Passionate Pilgrim: "She burnt with love, as straw with fire flamed. She burnt out love, as soon as straw out burnets" (cited in Enzmann & Kleiber, 1989).

The first few articles about burn out appeared in the mid-1970s in the United States (Freudenberger, 1974; Maslach, 1976). Initially, mainly practitioners and the general public were interested in burn out and the conceptual development was influenced by pragmatic rather than by scholarly concerns (Maslach & Schaufeli, 1993). The focus was on clinical descriptions of burn out.

The term "staff burn out" was first mentioned by Bradley (1969) in an article about probation officers who ran a community-based treatment programme for juvenile delinquents. Nevertheless Herbert Freudenberger

Drozdstoy St. Stoyanov (Ed.), New Model of Burn Out Syndrome: Towards Early Diagnosis and Prevention, 1–6.

(1974) is generally considered to be the founding father of the burn out syndrome (Maslach & Schaufeli, 1993). As an unpaid psychiatrist Freudenberger was employed in the St Mark's Free Clinic in New York's East Village for drug addicts that he was mainly staffed by young, idealistically motivated volunteers (Schaufeli & Buunk, 2003). Freudenberger observed that many of them experienced a gradual energy depletion and loss of motivation and commitment. This process took about a year and was accompanied by a variety of mental and physical symptoms. To label this particular state of exhaustion, Freudenberger chose a word that was being used colloquially to refer to the effects of chronic drug abuse: burn out. Freudenberger himself fall victim to burn out twice, which increased his credibility in spreading the message of burn out. His writings on the subject were strongly autobiographical and his impact is illustrated by the fact that in 1999, he received The Gold Medal Award for Life Achievement in the Practice of Psychology at the APA Convention in Boston (Schaufeli et al., 2009).

Independently and simultaneously, Christina Maslach became interested in the way people in the human services cope with emotional arousal on the job. She discovered that both the arousal and the strategies have important implications for people's professional identity and job behaviour. When by the chance she described these results to an attorney, she was told that poverty lawyers called this phenomenon burn out (Maslach & Schaufeli, 1993). Thus, a new expression was born.

But these facts do not indicate that the phenomenon burn out didn't exist before. Burish (1993) presents several examples of psychological states that have been described previously in the literature. The most illustrious example of burn out is the case-study of a disillusioned psychiatric nurse Miss Jones, published by Schwartz and Will (1953). Thomas Mann's description in Buddenbrooks (1901) includes the most essential features of burn out, such as extreme fatigue and the loss of idealism and passion for one's job. Graham Greene's novel A Burnt-Out Case (1960) tells the story of the world famous architect (a gloomy, spiritually tormented, cynical and disillusioned character) who quits his job and withdraws into the African jungle.

After the introduction of the concept by Freudenberger and by Maslach and her colleagues, burn out became a very popular topic in the next 5 years. More than 6000 books, chapters, dissertations, and journal articles on burn out had appeared (Schaufeli et al., 2009). This early literature had several noteworthy characteristics:

1. What was meant by the term burn out varied widely from one writer to the next;
2. The concept of burn out was expanded;
3. Burn out was largely non empirical. The most of early articles on burn out used a clinical approach (Maslach & Schaufeli, 1993).

During the next phase of the 1980s, the work on burn out entered a more focused, constructive, and empirical period. In the so-called empirical phase, seven trends may be observed:

- The introduction of short and easy to administer self-report questionnaire to assess burn out - Maslach Burn out Inventory (MBI);
- Burn out started to draw attention in countries outside the USA. In other countries burn out research started after the concept had been established in the USA, and after measurement instrument had been developed;
- Most research focused on people-oriented occupations. The publication of the MBI-General Survey allowed burn out to be studied independently from its specific job context;
- Research tended to focus more on job and organizational factors;
- Using longitudinal instead of cross-sectional designs in order to study the development of burn out over time;
- Traditionally most burn out research was rather theoretical;
- The concept of burn out is being supplemented and enlarged by the positive antithesis of job engagement (Schaufeli & Buunk, 2003).

By definition burn out is not considered as a mental disorder in the narrow sense. In the ICD-10 diagnostic system burn out (code 73.0) is placed in the category "problems related to life management difficulty" and described as "a state of vital exhaustion". Some authors (Acker, 2011) assume that the clinical burn out features overlaps with that of major depressive disorder based in the diagnostic criteria introduced in ICD-10 and coincides with *shift worker syndrome* which is relevant in particular for the health care professions.

According to the classical description of Ch. Maslach burn out includes three domains of clinical manifestations:

- Emotional exhaustion
- Depersonalization (in the sense of dehumanization)[1]
- Decrease in personal accomplishments

At present there is unanimity in the specialized literature about which professional groups are most likely to develop the Burn-out Syndrome. At

[1]In clinical psychopathology depersonalization is referred to as severe psychotic symptom

the same time, there is no standardized way of evaluating vulnerability to professional burn-out. The available standardized methods, such as Christina Maslach's burn out invcentory, can only account for irreversible functional changes in the individual after they have already taken place.

Our understanding is that burn out is complex and dynamic phenomenon both from clinical and management perspective. We adhere to the model of longitudinal course of burn out which is divided into three stages. The first one is **flame out** and it characterizes the effort of the subject to cope with the stressors of the working environment. We consider that the emotional exhaustion in the next stage, the **genuine burn out**, is rather a *consequence* from depressive and anxiety reactions during flame out stage. Finally when all personal resources are exhausted the irreversible stage of **rust out** is observed. In that stage the subject is completely alienated from others, cynical and radically ineffective.

In our paradigm it is possible to identify vulnerable groups of people for whom there are (Figure 1.1) approaches of preventing professional burn-out by psychosocial intervention. Based on the above stage model monitoring of the interaction between the psychological functioning of the individuals with the organizational climate in terms of psychological and social wellbeing should occur between the first two stages of the process – flame-out (i.e. stress of futile effort) and burn-out (i.e. exhaustion and demoralization). This early

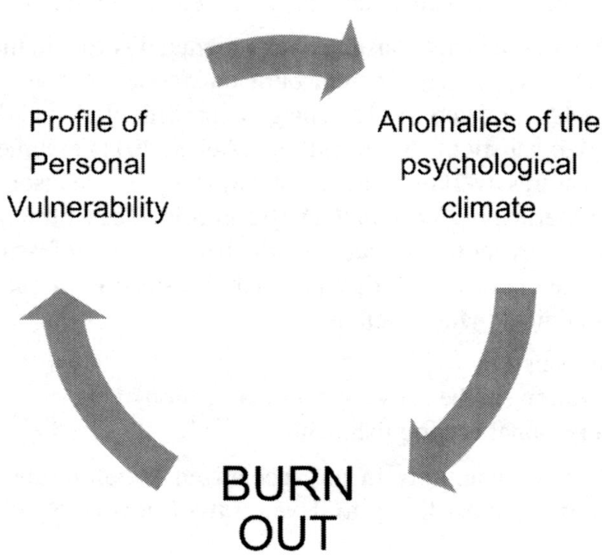

Figure 1.1 Model of complex vulnerability to burn out

period in the burning process is propitious to identify means of intervention for the sake of preservation and recovery of the psychological potential of the workers in vulnerable groups. At an earlier stage, the phenomenology of professional 'burning' is not difficult to differentiate from the symptoms of psychoemotional stress, whereas at a later stage (described by the term "rust out"), there may be less benefit from psychological interventions as the harm is thought to be irreversible. We are choosing to focus on early intervention, but hope that late interventions might also be helpful in future work (Stoyanov and Cloninger 2012).

We presume that the high-risk pattern which accounts for vulnerability to professional burning has three components. It includes characteristics of the interaction between the psychological and organizational climate that is combined with the personal traits of the workers. As a result, Burn-out Syndrome has three dimensions that can be measured and which are shown in Box 1 (Stoyanov and Cloninger 2012).

To summarize our concept the vulnerability to burn out is composed of diathesis (personality ***traits***) and stress (psychological climate at work place), complementary to specific vulnerable personality profiles.

This book presents the results from our comprehensive investigations in predetermined three high risk populations: heatlh and social care and penitentiary system.

Our model is tested most extensively in different sectors of health care system (psychiatry, intensive care units, etc.) and then compared with the vulnerability to burn out in social care. In addition anxiety and depression have been identified as sensitive ***state*** predictors of BOS and its impact on quality of life has been assessed

An independent study has been conducted in penitentiary system with different methods however under the same conceptual model of *stress-diathesis*

An appropriate preventive program is introduced in the final chapter.

References

[1] Acker, J, (2011). Health prevention of burn out and depression in the working environment, Turkish Journal of Psychiatry, Vol. 22, Issue 2, Suppl, 1: 42.

[2] Bradley, H.B. (1969). Community-based treatment for young adult offenders. Crime and Delinquency, 15: 359–370.

[3] Burisch, M. (1993). In search of a theory: some ruminations on the nature and etiology of burn-out. In W.B. Schaufeli, C. Maslach and T. Marek (Eds.). Professional Burn out: Recent Developments in Theory and Research. Washington, DC: Taylor & Francis.

[4] Enzmann, D., Kleiber, D. (1989). Helfer-Leiden: Stress und Burn out in psychosozialen Berufen [Helper Ordeals: Stress and Burn out in the Human Services]. Heidelberg: Asander Verlag.

[5] Freudenberger, H.J. (1974). Staff Burn out. Journal of Social Issues, 30: 159–165.

[6] Maslach, C. (1976). Burned-out. Human Behavior, 5: 16–22.

[7] Maslach, C., Schaufeli, W.B. (1993). Historical and conceptual development of burn out. In W.B. Schaufeli, C. Maslach and T. Marek (Eds.). Professional Burn out: Recent Developments in Theory and Research. Washington, DC: Taylor & Francis, 1–16.

[8] Schaufeli, W.B., Buunk, B.P. (2003). Burn out: an Overview of 25 Years of Research and Theorizing. In M.J. Schabracq, J.A.M. Winnubst and C.L. Cooper (Eds.), The Handbook of Work and Health Psychology. John Wiley & Sons, Ltd: 383–425.

[9] Schaufeli, W.B., Leiter, M.P., Maslach, C. (2009). Burn out: 35 years of research and practice. Career Development International, 14 (3): 204–220.

[10] Schwartz, M.S., Will, G.T. (1953). Low morale and mutual with drawal on a mental hospital ward. Psychiatry, 16: 337–353.

[11] Stoyanov, D. S., & Cloninger, C. R. (2012). Relation of People-centered public health and person-centered healthcare management: a case study to reduce burn-out. International Journal of Person Centered Medicine, 2(1), 90–95.

2

Theoretical Model of Vulnerability to Burn Out: Personality and Psychological Climate in the Context of BOS

Drozdstoj Stoyanov MD, PhD, Full Professor of Psychiatry, Medical Psychology and Person Centered Medicine,
Faculty of Medicine, MUP, Vice Chair Executive Committee PSIG, Royal College of Psychiatrists, Visiting Fellow, University of Pittsburgh

Boris Tilov PhD, Assistant Professor of General Psychology
Department of Health Care Management,
Faculty of Public Health, MUP

Bianka Tornyova PhD, Associate Professor of Medical Education,
Department of Health Care Management,
Faculty of Public Health, MUP

In order to disclose more precisely personality in the context of burnout, we should answer the question: What is temperament and character of a person and are thy relevant to the issues considered?

Temperament represents a common system of a procedure and joint existence of individual psychic features which determine the mental activity of man. The types of temperament are determined by the psychological characteristic of the features: Sensitivity, Reactivity, Activity, the ratio Reactivity-Activity, Rate of reactions, Plasticity and Rigidity. The classical typology of temperaments is based on an ancient classification of Hippocrates which describes four main human types according to 'the prevailing fluid in their organism'- sanguine (blood), choleric (yellow bile), phlegmatic (lymph) and melancholic (black bile).

Most psychologists define temperament by hereditary genotype characteristic for specific stable character traits, typical for the concrete person.

Drozdstoy St. Stoyanov (Ed.), New Model of Burn Out Syndrome: Towards Early Diagnosis and Prevention, 7–20.

In their study Thomas and Chess describe individual differences in babies, according to nine behavioural physiological- reactivity categories, as part of the overall rhythm, intensity of reactions, stubbornness and levels of activity (10). According to the authors from the earliest age specific individual behavioural models and styles, which influence the environment in which he/she resides, can be noticed and stand out (10).

Different understanding, interpretations and studies in the field of temperament exist. Witkin et al. assess and determine in a wide range of psychological systems temperament differences- perceptive, cognitive, emotional and social. Whereas Shapiro describes several neurotic styles according to experience and perception (10).

Temperament is a hereditary and more resistant characteristic connected with personality. According to Cloninger (4) temperament, as a basic essence of personality, includes 4 basic early- forming emotions – fear, aggression, affection and persistence of behavior. The emotions form as a response to external stimuli- danger: novelty, reward, frustrative nonreward, respectively. According to him the four temperament traits are unchangeable throughout a person's life due to their hereditary nature. In addition, they do not only reflect temperament essence but also the structure of personality disorders diagnosed on the basis of immaturity. [4 – 9].

Character is determined by the social reactivity of the personality.

Numerous researchers including Freud, Jung, From, Adler and Olport give different interpretations to the character typology. Adler and Olport oppose the general typologizing, considering that personality uniqueness cannot be defined by several meaningful combinations of words, but it is possible to make it precise if all those subtle shades of our mentality are deduced and terminologically stated. (10)

According to Bowbeck character is a regulator of the ego and impulses through specific values- attachment to the others, emotional devotion in clear aims, emotional control. He considers that in the presence of character disorder, we notice a deficit of these three characteristics (10).

Character peculiarities of the personality in mutual interrelation are referred to as *integral character portrait*. This portrait is built up on the basis of information connected with: age, social origin and status, communication, relationships; characteristics of the environment in which the person was brought up and grew up (physical and psychic aspect; parents).

Presently in the specialized literature, it is thoroughly studied that the healthcare sector, divided in the respective professional groups, is one of the most affected fields by the burnout syndrome. Meanwhile there is no standard model for assessment of susceptibility to professional burnout. The existing standard methods, such as the Christina Maslach test, only register irreversible changes that have already occurred in the functioning of the individual. On this background, the high levels of distress and professional burnout among medical specialties on state funding raise serious concern and research interest. The personality in its entirety and in the context of the syndrome is a prerequisite for elaboration on the dynamic nature of future prevention programmes. It is well-known that personality is built up in the process of our ongoing socialization; it is a dynamic structure, a set of knowledge, skills and specific experience.

The establishment of an eventual working model for a complex assessment of the psychoclimate in the healthcare institutions, which should be directed to an early diagnostics of burnout syndrome and its adequate prevention in accordance with personality, would provoke a necessary discussion for a change in the existing conservative system. According to us, this model should consist of three dimensions:

- Inductive assessment of the organizational socio-psychological climate which are subject to another study.
- Reduction in selfactualization as a key marker for increased susceptibility to professional burnout.
- Functional attitude towards the working environment according to the Katz model

The Cloninger test *(TCI - The Temperament and Character Inventory)* studies the character and temperament of the personality. The motivation for its application is connected with the hypothesis that selfactualization is not only affected in highest degree but also in the most early stages in burnout. One of the key scales of this test is designed precisely for study of self- transcendence, according to the terminology of the author himself.

2.1 Cloninger's Test-theoretical Foundations

The Cloninger's test is based on the psychobiological theory of personality (Cloninger, 1991). According to it, the personality structure is composed of two basic domains: temperament and character. Temperament according to Cloninger is biologically and genetically determined. In this line of thinking are also earlier studies, conducted by the author, as the unique Stockholm study of the antisocial behavior of adopted children. Cloninger and Boman prove the

significant benefit of genetic factors in the genesis of dependence and criminal behavior of children and adolescents. In this respect the psychobiological theory approximates Hank Eysenck's theory of human personality.
Temperament scales of TCI include the following elements:

- harm avoidance
- novelty seeking
- reward dependence

The other fundamental domain in the psychobiological theory describes the humanistic and trans-personal style of the personality. A specific accent in it is placed on the self- transcendence (revised self actualization according to Maslow). In this concept Cloninger seeks the key to the phenomenon of the psychosocial well-being. The corresponding scales from this domain include:

- self-directedness
- cooperativeness
- self-transcendence

Self-directedness is defined by the strong association with all indicators of health, including those for mental health (subjective satisfaction); as well as the social and physical health (as perception for health). Cooperativeness is assessed through social tolerance, empathy and readiness to ensure help, as well as the strong connection with social support, which exerts influence on well-being and the subjective experience of happiness. In its turn self-transcedence is defined through the powerful influence on the presence of positive emotions. It is measured through the capacity of the person to be engulfed by what he/she likes and to identify that which is beyond 'transient existence'.

In the described model it is the latter aspect from the functioning of the personality system that is most susceptible and impaired by the influence of the negative psychoclimate as a basic determinant for professional burnout in the highly differentiated professions including the medical one.

TCI (The Temperament and Character Inventory) is a tool, which according to Cloninger, encompasses 7 dimensions of personality (1, 2, 3). The questionnaire is developed to identify the differences between people in normal as well as in abnormal models of behaviour in seven main dimensions of temperament and character. By character he understands the automatic emotional reactions in different experiences, which are partially hereditary and remain relatively stable throughout one's life. The four assessed dimensions of temperament are:

1. Novelty seeking as an expression of behaviour activating system which determines activation of new reactions and signals for rewarding or avoiding as a punishment and apart from that, has to play a significant behaviour modulated role especially dopaminerge system. Patients with whom high levels of the curiosity scale are reported, according to Cloninger et al., are described as seeking, curious and impulsive (3). Patients with low levels are rather indifferent, thoughtful and modest.
2. Harm avoidance is an expression of the behavioural inhibition system which includes the reactions to signals for punishment or lack of reward and is dominated by the serotonergic system. Personalities who avoid harm are described as anxious, pessimistic and tentative. Low level of harm avoidance means a calm, optimistic and assertive behavior.
3. Reward dependence as an expression of the behavior maintenance system, which makes further behaviour possible, without allowing tension and in which norepinephrine system is of crucial importance. High values of reward dependence run concurrently with devotion, turning and dependence. Patients with low degrees are described as purposeful, cold and detached (individualistic).
4. Persistence. High values of persistence denote that the people are diligent, persistent and perfectionists. The low values of this scale correspond to passive, disoriented behavior without pretentions.

Character is based on the individual concept and differences in aims and values, which exert influence on freedom in making decisions (freedom of choice), intentions and meaning of what a person experiences (life experience). To a certain degree the character differences are influenced by the socio-cultural life and constantly manifest themselves throughout the life.
The three dimensions of character are:

1. Self –directedness. Patients with high self-directedness are defined as responsible, purposeful and are able to integrate well (flexible). Patients with low results in this scale are ineffective, lay the fault on others (blame others) and are too indecisive and clumsy (slow and awkward).
2. Cooperativeness. Patients designated as cooperative are responsive compassionate, creative and socially tolerant. Low levels reveal vengeful, destructive and unprincipled men.
3. Self-transcendence. High values of self-transcendence are connected with patience and experience and are recorded in creative and completely dedicated individuals. Low values indicate lack of humility.

2.2 Well-being

Well-being encompasses a variety of subjects and is an object of study in many disciplines such as psychology, philosophy, sociology. A subject of study of well-being is, on the one hand, the concept for subjective well-being, and on the other, psychological well-being. It is part of the positive psychology whose founder is Martin E. P. Seligman (11). As a starting position, he names there basic tasks of the psychology in the period before the Second World War: 'To treat mental disorders, to organize more productive and more valuable existence for all people, to recognize and encourage talent "(12). After World War II the focus mainly shifts onto mental disorders and their treatment while the other two tasks fall into oblivion (12).In the last years the resources, powers, respectively skills and values, as well as the quality of life are one more object of attention and prevention of the disease acquire significant importance. The goal of positive psychology is described by Seligman and Csikszentmihalyi (2000):

"The aim of positive psychology is to begin to catalyze a change in the focus of psychology from preoccupation only with repairing the worst things in life to also building positive qualities." (p. 5)

Positive psychology is divided into three aspects (12): well-being and satisfaction (In the past), hope and optimism (for the future), flow and happiness (in the present).

On the individual level, the discussion is about positive characteristics, as, for instance, the ability to love, talent, courage, the ability for interpersonal communication, aesthetics, persistency and tenacity, forgiveness, self-confidence, future mindedness, gifts, spirituality and wisdom.

On the group level it concerns social values and institutions, the individuals to be better physical citizens, which refers to responsibility, caring compassion and nuturance, altruism, politeness, moderation, tolerance and professional ethics.

In many definitions for health, well-being is a key component. As the definition of WHO indicates: 'Health is a condition of complete physical, mental and social well-being, and not just lack of disease or ailment. The best health condition is one of the basic rights of every human being, without difference in race, religion, political views, economical or social conditions." (WHO, 1946)

This definition does not remain without criticism since it originates from the permanent state of absolute well-being which verges more with utopia, than with real life (13). What is significant in this definition is consideration of well-being in several aspects: physical, psychological and social. Wydra depicts this as a multidimensionality of the socially-oriented health condition which stands out against the physiologically-oriented medical term and fins expression in the WHO definition.

2.2.1 Mental Well-being

Some authors divide well-being into mental and subjective (14, 15), others perceive mental and subjective well-being as two aspects of one and the same construct (16), and respectively consider the subjective as part of the mental well-being (17, 15).

According to Keyes et al. (2002) mental well-being refers to incessant solving of existential problems and challenges of life, in other words, the pursuit and achievement of important goals such as growth and development of personality and creation of qualities for relation, attitude, commitment, connecting link with the others.

Carol Ryff (18) differentiates among six dimensions or qualities of mental well-being: self-acceptance, positive relationships with the others, autonomy, control of the environment, existence with goals and personal growth.

2.2.2 Subjective Well-being

The idea of subjective well-being originates from Greek hedonism (19). There are many definitions of subjective well-being in literature. On the one hand, they are partially used as synonyms of the concepts well-being, happiness and satisfaction, on the other hand, the necessity of strict distinction between the two is emphasized.

As early as 1967, on the basis of his study on happiness, Wilson defines the happy person as 'a young, healthy, well-educated, well-paid, extroverted, optimistic, worry-free, religious, married person with high self-esteem, job morale, modest aspirations, of either sex and of a wide range of intelligence" (20).

On the basis of this definition a number of studies have been conducted. Although many of the conclusions of Wilson are outdated- youth and modest expectations are not perceived any more as a prerequisite for well-being and happiness- still some of his conclusions are repeatedly examined and studied (20).

Diener at al. (20) define subjective well-being as: "... a broad category of phenomena that includes people's emotional responses, domain satisfactions, and global judgments of life satisfaction."

Diener and Tov (19) point out that subjective well-being is associated with happiness. Even Wydra (22) speaks of the usage of the terms happiness and well-being as synonyms. It is important to mention that in using the term we should proceed from the subjective assessment (21).

Mayring (1991) defines happiness as a positive, long-term, emotional, cognitive factor for well-being.

The component satisfaction from life is to be defined as lack of complaints, loading and worries and having certain means and given standard of life, as well as the individual quality of life, although not only the objective conditions of life have significance, but also the subjective assessment for them (22). On the one hand, the affair in question is satisfaction from life as a whole, and on the other, satisfaction from separate fields of life such as partnership, marriage, family life and profession (23).

Well-being occupies a central position when considering quality of life. In his book "Feeling Good: The Science of Well- Being", Cloninger [24] presents a complete concept of the subjective well-being and reveals a manner for fulfilling existence. Personal well-being is of crucial importance for the manner in which the other dimensions of the quality of life would be assessed, such as social integration, social and material conditions of life etc.

2.2.3 Significance of Subjective Well-being

Subjective well-being is a significant indicator not only for the quality of life and health of a particular person but also for the quality of a whole society. According to Diener, Oishi und Lucas (20), good life, the happy person and happy people would create a happy and good society. Although people's well-being in itself does not create a good society, it is an essential and necessary part. The individual, as well as politics seem to seek always good and happy life. (25) Well-being, happiness and satisfaction are significant goals of life and specific strategies for their implementation and achievement more and more frequently find application in therapy and consultations (21).

A number of studies indicate the positive consequences of happy life: strengthening physical and mental health, more active attitude to life, high level of realization, sensitivity and openness to reality, empathic social orientation and integrating and supportive influence on personality (21).

Well-being, happiness and satisfaction have their significance in many different fields (21): in philosophy, which deals mainly with happiness, in Christian theology, in the theory of literature, in the economic and social sciences (well-being as a social factor, as an indicator for quality of life, as an object of study in national and international research, in gerontology (happiness as an indicator for successful ageing), in psychology (for example, psychology of emotions and psychology of healthcare) and in physiology (physical foundations of well-being).

2.3 Psychological Climate

Psychological climate in its essence has different manifestations in the healthcare structures. Increasingly, in the developed and socially responsible countries, attention is paid to the psychological aspects which presuppose the development of a favorable social environment for interaction in the workplace of the healthcare institutions. This is due to the essence of the work and the threats of the environment which would exert influence on the employees, through professional exhaustion, as well as on the patients through the attitude toward them which is not that humane and careful.

Psychosocial climate is determined by a number of internal and external factors and is characterized in different sources such as: climate of interaction, favourable or infavourable atmosphere, safe work environment, team climate, stress resistant management strategies, group solidarity. The moment of shared interaction of the members of the group or team may not be directly connected to the mission and properties of the given organization (26). According to Levin, the climate of the group is a consequence of shared perceptions among the people or grazing of the individual cognitive maps (26). This mutual atmosphere may be characterized as levels and degree of support of leaders for employees, satisfaction through adequate remuneration and incentive, to what degree there is freedom of choice in group and individual decisions, precise informing on the goals and mission of the organization, how conflicts are solved, feeling of usefulness, appurtenance and activity of the separate employee, method for distribution of tasks (26).

Undoubtedly all this would not be realistic without adequate leadership approach, management style which would direct the trend of development of the whole organization, as well as of every single employee in it. The other factors are formal systems, structures, administrative rules and procedures for specific decisions (26). The research on the topic prove that the psychosocial climate influences fully motivation, satisfaction and work or in other words, the positive climate is a good prevention from emotional and professional burnout.

A study of A. Petkov et al. on the influence of the work environment on the medical staff reveals that it is the psychosocial climate that is defined as the most important of all criteria. 94% of the surveyed indicate that the 'microclimate of the work place is' the most important for them. 70% or the second most important factor are 'encouragement and cooperation from the leader' and 'contact and team work with colleagues'. As a following aspect with 65% is defined the answer 'acknowledgement for the work done'(28).

In his article "Measuring organizational culture and climate in the health-care', Robin Gershon and Patricia Stone, register the necessity from a similar tool on the basis of specific conclusions from medical practice and its administration. They describe the foundation of terminological definition of psychological climate and culture from the beginning of 1930s till its historical development in 1980s.

After a specific research in 2001 of 21 studies for measuring the patient's safety, it becomes evident that 2/3 of results are related to organizational climate and organizational culture. These data justify the establishment of clear, accurate and specific valid measures for the organizational functioning of the healthcare institutions. The aim of these studies is to clarify, on the one hand, the definition for organizational culture and climate and a process of standardization of terminological differences to begin. In the mean time specific instruments, which measure the organizational climate and culture, should be determined.

The authors make an extensive overview of the existing methods on the topic by dividing them into such that are related to healthcare and others which are not. Subsequently they realize the ineffectiveness of the approach and generalize the existing tools under several points: 1) full citation of the original article; 2) assessment of constructions and sub constructions; 3) psychometric properties of subconstructions, relying only on one type of validation which includes additional psychometric tests for factor analysis; 4) initial aim and aim of the tool; 5) Full citation of every article which refers to the authentic references if there are such; 6) Summarization of the results related to healthcare. The choice of resources reaches to the categorization of a specific psychometric approach, combined with citation in scientific communities.

After the initial search 311 sources were found, only 12 of them encompass the last 20 years, with correct psychometric approach on the searched criteria. Most authors report that only validity is a consequence of correlational analysis.

The authors identify from 10 instruments 116 different subcategories which afterwards they arrange in 4 basic groups:

1) Leader style such as degree and type of management, degree of support and trust, degree of isolation and type of management hierarchy.
2) Group behavior and relationships which includes perception of trust for the colleagues, degree of group support, group cohesion and coordination of group activities.
3) Communication (for example, formal and informal mechanisms for information transfer and conflict resolution);
4) Structural characteristics of the professional life (for example, payment, work conditions, working time, forced overtime work, occupational safety);

The authors point out, identifying the leading results related to healthcare, as most common patient satisfaction, job satisfaction, motivation, stress at work and fluctuation of staff (number of staff who joined an left the organization).

Another essential point of the organizational interactions or part of the overall climate is considered the safe climate in its management, social and interpersonal aspect. Safety is a product of the culture of interaction, of the individual and group aspects, world outlook, perception, management competence on the personal engagement, style and proficiency.

Organizations which are in the high-risk category should regularly make an assessment of their culture of safety. This is a product of the individual and group values, attitudes, perceptions, competences, models of behavior, style and experience or management of safety in the organization. This approach may be implemented since all organizations have hierarchical structure and are created on different levels on which the climate can, with certainty, be studied. Including healthcare organizations. The tools used aid in making an accurate assessment of the organizational climate. The result of this information often serves managers as an additional perspective concerning the safety culture.

The most commonly used indicators for assessment of climate safety in the health care are: Managers, Safety systems, Risk perception, Work requirements, Reporting, Safety relationships, Communication (feedback), team work, Personal values, Resources, Organizational factors.

In relation to the study of safety climate in healthcare organizations, several tools have been developed. For this purpose, the criterion for encompassing and detailed consideration has been determined:

1) To use a questionnaire for individual responses, especially designed to assess the climate for safety or culture of safety in the healthcare institutions;
1) Details related to the correct use of data and assessment of the tools.
2) More than 50 participants should be tested.
3) The report should be published in the corresponding language, where it has been conducted.

For a research to be assessed, psychometric criteria are established:

1) Content validity
2) Relative validity
3) Factor analysis

This aspect is important to us due to the necessity of designing working and effective programmes which would ensure the well-being of the healthcare professionals and from there, as a reflection, on the patients themselves. One such approach would prove a successful management strategy which would preserve the positive atmosphere and psychoemotional state of each of the members of the group or team in the healthcare organization.

References

[1] Cloninger CR (1986) A unified biosocial model of personality and ist role in the development of anxiety states. Psychiatric Developments 3: 167–226;

[2] Cloninger CR (1988) A unified biosocial theory of personality and ist role in the development of anxiety states: A reply to commentaries. Psychiatric Developments 2: 83–120;

[3] Cloninger CR, Svrakic DM; Przybeck; TR (1993) A Psychobiological Model of Temperament and Character. Archives of General Psychiatry 50: 975–990).

[4] Cloninger CR, Svrakic DM, Przybeck TR. A psychobiological model of temperament and character. Arch Gen Psychiatry 1993;50: 975–90.

[5] Cloninger CR, Przybeck TR, Svrakic DM, Wetzel RD. The Temperament and Character Inventory—a guide to its development and use. St. Louis: Washington University; 1994.

[6] Cloninger CR, Przybeck TR, Svrakic DM, Wetzel RD. The Temperament and Character Inventory—Revised. St. Louis: Washington University; 1999.;

[7] Cloninger CR, Svrakic DM. Personality disorder. In: Sadock V, Sadock B, editors. Comprehensive textbook of psychiatry. 9th ed. Williams and Wilkins; 2009. p. 2197–240.

[8] Dzamonja-Ignjatovic T., Cloninger R., et all. Cross-cultural validation of the revised Temperament and Character Inventory: Serbian data. Compr Psychiatry 2010; 50: 649–655

[9] Miettunen J, et al. International comparison of Cloninger's temperament dimensions. Pers Indiv Diff 2006;41(2006): 1515–26.

[10] Kreyghead, E., Nemerof, Ch. Encyclopaedia in nPsychology and Behavioural Science, Science and Art, Sofia, 2008

[11] Martin E. P. Seligman (2002). Positive Psychology, Positive Prevention, and Positive Therapy. In C. R. Snyder (Ed.), Handbook of positive psychology (pp. 3–9). Oxford: Oxford

[12] Seligman, M. & Csikszentmihalyi, M. (2000). Positive Psychology. An Introduction. American Psychologist, 55(1), 5–14

[13] Wydra, G. (2005). Der Fragebogen zum allgemeinen habituellen Wohlbefinden. Saarbrücken: Sportwissenschaftliches Institut der Universität des Saarlandes.

[14] Emmons, R. A. (1992). Abstract versus concrete goals: Personal striving level, physical illness, and psychological well-being. Journal of Personality and Social Psychology, 62(2), 292–300

[15] Keyes, C. L. M., Shmotkin, D, Ryff, C. D. (2002). Optimizing Well-Being: The Empirical Encounter of Two Traditions. Journal of Personality and Social Psychology, 82(6), 1007–1022

[16] Ryan, R. M. & Deci, E. L. (2001). On happiness and human potentials: A review of research on hedonic and eudaimonic well-being. Annual Review of Psychology, 52, 141–166

[17] Diener, E., Suh, E. & Oishi, S. (1997). Recent Findings on Subjective Well-Being. Indian Journal of Clinical Psychology, 24, 25–41

[18] Ryff, C. D. (1989). Happiness is Everything, or Is It? Explorations on the Meaning of Psychological Well-Being. Journal of Personality and Social Psychology, 57(6), 1069–1081

[19] Diener, E. & Tov, W. (2006). National Accounts of Well-Being. In K. C. Land (Ed.), Encyclopedia of Social Indicators and Quality-of-Life Studies (in press).New York: Springer

[20] Diener, E., Suh, E., Lucas, R. & Smith, H. (1999). Subjective Well-Being: Three Decades of Progress. Psychological Bulletin, 125(2), 276–302

[21] Mayring, P. (1991). Psychologie des Glücks. Stuttgart, Berlin, Köln: Kohlhammer

[22] Wydra, G. (2005). Der Fragebogen zum allgemeinen habituellen Wohlbefinden. Saarbrücken: Sportwissenschaftliches Institut der Universität des Saarlandes

[23] Diener, E. (1994). Assessing subjective well-being: Progress and opportunities. Social Indicators Research, 31(2), 103–157

[24] Cloninger R. Feeling Good: The Science of Well- Being. Oxford, New York, 2004

[25] Veenhoven, R. (1994). Is happiness a trait? Tests of the theory that a better society does not make people any happier. Social Indicators Research, 32(2),101–160.

[26] Крейгхед, Е., Немероф, Ч., Енциклопедия по Психология и Поведенческа наука, Наука и изкуство, София, 2008

[27] Gershon, RRM, Patricia W. Stone, Suzanne Bakken, Elaine Larson, Measurement of Organizational Culture and Climate in Healthcare, JONA, Volume 34, Number 1, pp 33–40 2004

[28] Petkov A., et al., Influence of the work environment on the activity of nurses. Social medicine 2, 1997, p. 35–39

3

Empirical Measures of Vulnerability to BOS. Standardization and Validation of the Battery of Assessment Tools

Ralitsa Raycheva[1] Assistant Professor,
Department of Social Medicine and Public Health,
Faculty of Public Health, MUP

In modern health-care systems, well-being and burnout are becoming increasingly relevant topics. The professional groups most affected by the burnout syndrome in health-care system are well defined in the literature. However, there is no standardized method for establishing vulnerability to burnout.

The factors associated with organizational behavior in evidence based management practice are most often assessed using standardized tools allowing systematic monitoring and comparative analyses across various organizational environments.

The use of linguistically validated scales could resolve the important problem of a lack of tools available in non-English countries. However, this approach raises methodological problems of cross-cultural adaptation resulting from variability of organizational and cultural factors.

Standardized research instruments facilitate data collection and monitoring, yet standardization requires that research instruments be adapted for use in different cultures because of the increase in diversity in research that examines practices across different countries.

The major challenge is not as simple as translating an instrument into a different language but addressing specific properties when selecting and adapting research instruments such as psychometric properties (validity, appropriateness, reliability and responsiveness) as well as feasibility and acceptability of the instrument.

[1]The contribution to this part of the research program of *Radost Assenova, Dimitar Kazakov, Simeon Yordanov, Tanya Turnovska, Nonka Mateva, Drozdstoy Stoyanov* is acknowledged

Drozdstoy St. Stoyanov (Ed.), New Model of Burn Out Syndrome: Towards Early Diagnosis and Prevention, 21–46.

Translation is the most common method of preparing instruments for cross-cultural research and has pitfalls that threaten validity. Some of these problems are difficult to detect and may have a detrimental effect on the study results. Identification and correction of problems can enhance research quality and validity [1]. Evidence must be provided that the meaning of items in the translated version is equivalent to items in the original language.

Translation of concepts and categories is usually preferable to *verbatim* translation of all data transcripts [2].

A multistep process of preparation, forward translation/ reconciliation, back translation/back translation review, harmonization, cognitive debriefing, review of cognitive debriefing results and finalization is often proposed to ensure that certain requirements are met [1, 3]. Commonly referred to in this context are five categories of equivalence:

- **Content Equivalence**. The content of each item of the instrument is relevant to the phenomena within each culture studied.
- **Semantic Equivalence**. The meaning of each item is the same in each culture after translation into the language and idiom (written or oral) of each culture.
- **Technical Equivalence**. The method of assessment (e.g. pencil and paper, interview) is comparable in each culture with respect to the data it yields.
- **Criterion Equivalence**. The interpretation of the measured variable remains the same with respect to the norm within each culture.
- **Conceptual Equivalence**. The instrument is measuring the same theoretical construct in each culture.

Statistical methods used for evaluating cross-cultural equivalence include correlation between alternate forms administered to bilinguals to assess content equivalence and item-matching, differential item functioning (technical equivalence), regression analyses (criterion equivalence), correlation methods, factor analysis for conceptual equivalence, etc.

The psychobiological and organizational profiling of psychosocial well-being required the use for the first time of a battery of tests - Cloninger's test (TCI-R; permission courtesy of C.R. Cloninger) and then standardized tests of Maslach, and Koys and DeCotiis as well as content analyses of organizational documents containing data on turnover, leaves of absence etc. to test the hypothesis of relation between burnout as a factor and organizational indicators as dependent variables.

3.1 Measures

3.1.1 Temperament and Character Inventory

The temperament and character inventory (TCI; Cloninger *et al.*) [4] is a self-report questionnaire developed to assess the seven dimensions of personality described by Cloninger *et al.* in 1994, with a total of 29 subscales [4]. In 1999, a revised version was proposed by Cloninger (TCI–R) [5]. The response format of TCI-R increases the reliability of the responses employing a 5-point Likert scale (1 - definitely false to 5- definitely true).

In common with most instruments in psychology, TCI was developed with regard to a particular population, in this case, the USA. Currently, much research focuses on the reliability and validation of TCI in various populations, including those of the Dutch, English, French, Korean, Swedish, Polish and Turkish [6–14].

The Cloninger test was standardized for the Bulgarian population by K. Kalinov, V. Milanova and A. Jablenski in the Spring of 2003, with the results published in 2005. [15] The structure of the test according to that standardization meets the dimensions specified by Cloninger in the American population. In a sample of 843 healthy individuals studied with the assistance of NORC in April 2003, high internal consistency, reliability and validity of the applied scales were found. One year after the standardization of Kalinov et al., Cloninger subjected his method to fundamental revision.

The revised TCI-R (2004) has not been applied in the Bulgarian population. Moreover, neither the original version of the test, nor its later revision have been used in our “target” population, namely a sample of physicians and other health professionals at risk of burn out syndrome.

The applicability of the questionnaire in many European countries as well as previous work on standardization of the earlier unrevised version of TCI for the Bulgarian population gave us grounds to consider its integration into a set of instruments along with two other tools – IMPC. MBI in the Project studying specific target groups such as the medical professions in Bulgaria.

The validation process results are also anticipated to contribute to current research studying the psychometric properties of the TCI-R in different societies, cultures and social environments [7–17].

3.1.2 Inductive Measurement of Psychological Climate (IMPC)

The test of Koys and DeCotiis [18] is also a self-report measure with 8 categories of psychological climate perception: Cohesion, Trust, Support, Fairness, Autonomy, Pressure, Recognition and Innovation. In this study, we used 40 items describing the organizational climate and a 5-grade Likert scale to comply with the scale used in the TCI-R section of the complied tool. Participants were asked to rate how each description corresponded to their own perception of the working environment.

3.1.3 Maslach Burnout Inventory (MBI)

MBI weighs the effects of emotional exhaustion and reduced sense of personal accomplishment. It consists of 22 items yielding scores for three components.[19–20] Emotional exhaustion (EE) - defined by the depletion of energetic resources to "give" to others is very specific to the medical professions.

In the Bulgarian standardization of Maslach's test, performed in 1992 by B. Tsenova, the Depersonalization is translated in a broader sense as "dehumanization" and reduced Personal accomplishment (PA), which consists of reduced interest in professional development and improvement, personal achievements have been reformulated as "work efficiency".

We used the original 7-grade scale for each item (0-never to 6-always).

The proper application of the battery comprising the three tools enables the analysis of the consequences of the professional burning process along with analysis of the psycho-emotional environmental effects.

Since two of the inventories - IMPC and TCI-R had not been validated in Bulgarian the initial stage of the Project aimed to provide Bulgarian versions and to test the conceptual validity and cross-cultural applicability of the compiled tool in a culturally, socially and economically diverse settings.

3.1.4 Linguistic Validation of the Bulgarian Version of the Set of Tools

The TCI-R, initially designed in English, was translated into the Bulgarian language by four translators following the recommended stages of translation and cultural adaptation: translation with conceptual and linguistic evaluation, back translation, comparison of the source and target versions, and verification of the new instrument.

Two native target-language-speaking translators, bilingual in the source language, produced a forward translation of the original items, instructions and response choices. A single combined version in a colloquial and easy-to-understand language was prepared by the translation process manager, discussed and agreed on by the expert research team. The process involved changes with respect to the source language version, where words or concepts were untranslatable or where words or terms have a specific meaning in one language but a semantically different or secondary meaning in the other language.

The first reconciled forward version of the questionnaire was next translated back into the source language (English) by bilingual professionals unfamiliar with the original tool.

Comparison of the backward version with the original source version was performed by the project manager (expert in psychology) during a meeting with the expert supervising the translations in order to detect any misunderstandings, mistranslations or inaccuracies in the intermediary forward version of the questionnaire items, instructions and answers.

As a result items were marked as 'matching', 'very close' or 'acceptable' in meaning, wording and grammar. Overall 10 items have been revised because of differences in meaning, inadequate wording or typographical errors.

The final version of the battery was proofread by an independent native target-language speaker in order to perform a final check of the spelling, grammar, page layout, etc., and administered to a panel of five respondents to determine whether the translation (instructions, items and response choices) was acceptable, whether it was understood in the way it was supposed to be, and whether the language used was simple and appropriate. Two main aspects were tested:

- the conceptual equivalence of items and response choices/ratings in the translation with the original; and
- the respondents' understanding of the instructions and items, and any problems in the administration of the translated questionnaire.

The revised and agreed version of the tool, accompanied by detailed reports of the overall process containing comments and notes outlining the translation issues that were discussed, and how the final decisions were made, were submitted for review by prof. Cloninger. The approval indicated that introduction and instructions did not differ significantly and neither did the items and answer scales used.

The linguistic validation process was much more time consuming than a single straightforward translation, taking about 3 months to accomplish, but the team work and the development of a consensus and the translation by committee methodology reduces the cultural and social bias that may result when only one or two translators are responsible for the translation.

3.1.5 Study Design

Two studies were performed to pilot run and assess the quality of the compiled tool and to gather data from a larger sample of the target professional groups at risk of burnout.

The subjects were recruited from health-care and social establishments. All subjects consented to participate in the studies.

Both pilot and field study used the agreed Bulgarian version of the tool comprising TCI-R, IMPC, MBI and their results were used in validity and reliability assessment presented here.

3.1.6 Test–retest Study

The aim of administering the tools to a pilot target population sample was to determine the equivalence of the source instrument with the newly translated target-language version through item-level and scale-level analyses.

The test–retest interval was 2 months ± 2 weeks. The sample consisted of 35 subjects, between 25 and 65 years of age. All professional positions and specialties were included to match the normative sample of health professions at high risk of burnout as closely as possible. Most of the respondents were doctors (19, 54.29%), 20.00% (7) were midwifes, 11.43% (4) were nurses, and the remaining were laboratory assistants (2, 5.71%) and a rehabilitation therapist (1, 2.86%). A total of 11 (31.43%) subjects were from surgical and 24 (68.57%) from other therapeutic specialties, which is roughly paralleling the ratio of professionals' distribution in the health-care system in Bulgaria. Some of the participants were health-care professionals from specialties other than the earlier mentioned as mostly affected by burnout syndrome. (Table 3.1)

3.1.7 Field Study

Despite the targeted larger sample in the field study non-reponse and feasibility issues (length of the questionnaire, time required to fill it, inadequate organizational involvement of health-establishment authorities etc.) accounted for the total of 302 questionnaires returned. Respondents' demographics did not differ significantly from the participants in the pilot run although the sample included more professional positions and a group of social workers that were used as a reference in further analyses of burnout presented in other chapters. (Table 3.1)

Table 3.1 Demographics and professional status of participants in pilot and field study

Demographics/Professional characteristics	Pilot study	Field study
	N (%)	N (%)
Female	23 (65.7)	209 (69.2)
Males	12 (34.3)	93 (30.8)
	mean ± SD	**mean ± SD**
Age	43.89 ± 9.24	43.26 ± 10.69
Work experience	17.57 ± 9.54	18.71 ± 11.09
Specialty	**N (%)**	**N (%)**
oncology	0 (0)	4 (1.3)
surgery / ER	9 (25.7)	101 (33.4)
internal dis / GPs	9 (25.7)	66 (21.9)
psychiatry	7 (20.0)	25 (8.3)
other	10 (28.6)	77 (25.5)
social workers	0 (0)	29 (9.6)

3.1.8 Statistical Analyses

The data was also used to determine mean score and standard deviation values for the subscales. Likert scale variables were treated like interval to allow the use of common parametric tests. [21] Missing values (<0.01%) were excluded.

A set of statistical analyses traditionally used in cross-cultural studies include internal consistency of the scales using the coefficient of Cronbach (Cronbach's alpha). Reliability was also assessed by interpreting the item-total correlations in subscales. The criteria used to identify non-homogenous items was correlation of <0.25 between the item and the subscale score.

A test–retest analysis of the scale scores was performed in the pilot sample ($n = 35$), and Pearson's correlation was calculated for the two sets of responses to our battery of psychological tests.

Test–retest correlations and interscale correlation coefficients for the subscales were also compared with data reported in other studies using the original and translated versions of the questionnaire.

Battery reliability was analysed using regression analyses and structural modelling and the results are presented in a separate chapter.

Statistical processing was performed using Statistical Package for Social Science (SPSS v.17).

3.1.9 Pilot Run Results

There were strong correlations between the results obtained from the two runs of the pilot study except for one subscale (SD3 resourcefulness) with medium correlation of the test and retest scores. There is no significant difference between males' and females' scores of TCI-R, IMPC, MBI and the scales did not correlate highly with age and work experience.

The internal consistency of the scales was tested as a general measure of its construct validity. Cronbach's alpha coefficients of TCI-R, IMPC and MBI scales are all satisfactory. As shown in Table 3.2, with few exceptions most alphas exceed 0.70. The coefficients in both test and retest ranged from 0.75 to 0.96 for the tools included in the battery indicating high level of reliability by the accepted criteria. Highest levels of reliability were established for the IMPC scales with α values about 0.95 in both measurements except for the Pressure scale.

The results confirmed the reliability of TCI-R with α values approximately 0.8 for all scales (0.9 for SD). The subscales had somewhat lower but still acceptable internal consistency comparable with data in other studies [4, 5, 16, 18]. Our pilot results on four subscales with α less than 0.50 [harm-avoidance (HA3 shyness), cooperativeness (CO2 empathy, CO3 helpfulness), RD4 dependence and SD3 resourcefulness] are consistent with findings in a study using TCI in another post-communist, South-eastern and Slavonic country on the Balkans – Serbia. The cultural similarities and background may account for the differences in TCI scores and their determinants compared to the American population that the tool was initially designed for [16, 22].

Our pilot results supported the cross-cultural applicability and the construct validity of the compiled battery of tools based on the strong correlations of subscales scores in the test–retest study (r = 0.63–0.89) except for the subscales with low α.

Base on the results from the pilot test for standard deviations in the subscales and calculating for 95% probability and 5% error we determined the needed number of participants for the field study to be 475. However even after adjusting for expected non-response the return rate was low resulting in a total of 302 questionnaires finally processed. The reasons for the non-response are to be elaborated further but eventually led to the adoption of alternative approaches such as on-line web-based questionnaires.

3.1.10 Field Test Results

The analyses of the items and their relation to the tool reliability in the larger sample results indicated that in some subscales of TCI-R there were questions that did not satisfy the homogeneity requirements.

Identifying inconsistent items was based on the following criteria – increase of more than 0.10 in the total scale reliability (Cronbach α) if item deleted or corrected item to total scale score correlation less than 0.25.

Table 3.2 Internal consistency (Cronbach α) of the copiled tool in test-retest study

Scales	Temperament scales	Test (n = 35) No. of items	Mean	± Standard devaition	Cronbach's α	Retest (n = 35) Mean	± Standard devaition	Cronbach's α	Test – retest correlation coefficient
		240			0,851			0,840	
NS	**Novelty-seeking**	35	100,23	11,543	**0,751**	100,18	12,869	**0,810**	0,795***
NS1	Exploratory excitability vs rigidity	10	31,71	4,219	0,559	30,56	4,405	0,569	0,543**
NS2	Impulsiveness vs reflection	9	23,94	3,955	0,558	25,32	4,290	0,613	0,649**
NS3	Extravagance vs reserve	9	26,31	5,524	0,735	26,50	4,907	0,633	0,775**
NS4	Disorderliness vs regimentation	7			0,531	17,79	3,514	0,536	0,705*
HA	**Harm-avoidance**	33	90,60	14,294	**0,885**	87,97	13,608	**0,880**	0,690**
HA1	Anticipatory worry vs optimism	11	30,86	4,821	0,678	29,26	5,148	0,754	0,606**
HA2	Fear of uncertainty vs confidence	7	22,06	4,582	0,742	21,44	4,301	0,680	0,626**
HA3	Shyness vs gregariousness	7	18,74	3,062	0,455	17,82	3,316	0,544	0,731**
HA4	Fatigability and asthenia vs vigor	8	20,26	4,749	0,805	19,44	4,265	0,756	0,810**

Continued

Table 3.2 Continued

		Test (n = 35)				Retest (n = 35)			
Scales	Temperament scales	No. of items	Mean	± Standard devaition	Cronbach's α	Mean	± Standard devaition	Cronbach's α	Test – retest correlation coefficient
RD	**Reward-dependence**	30	100,29	12,109	**0.835**	98,38	10,404	**0,774**	0,729**
RD1	Sentimentality vs insensitivity	8	29,09	3, 966	0,695	28,24	3,619	0,600	0,562**
RD2	Openness to warm communication vs aloofness	10	34,06	5,461	0,752	33,18	4,435	0,549	0,629**
RD3	Attachment vs detachment	6	19,63	4,815	0,810	19,53	4,129	0,779	0,814**
RD4	Dependence vs independence	6	17,51	2,228	0,242	17,44	2,338	0,062	0,483**
PS	**Persistence**	35	123,14	14,471	**0,888**	120,71	15,416	**0,899**	0,867**
PS1	Eagerness of effort vs laziness	9	30,94	4,814	0,738	30,94	4,177	0,664	0,824**
PS2	Work hardened vs spoiled	8	28,40	3,920	0,709	27,76	4,222	0,763	0,725**
PS3	Ambitious vs underachieving	10	35,94	5,011	0,745	34,76	5,428	0,774	0,861**
PS4	Perfectionism vs pragmatism	8	27,84	3,582	0,534	27,24	4,250	0,613	0,808**

Scales	Temperament scales	Test (n = 35) No. of items	Mean	± Standard devaition	Cronbach's α	Retest (n = 35) Mean	± Standard devaition	Cronbach's α	Test – retest correlation coefficient
SD	Selfdirectedness	40	140,06	18,812	**0,911**	140,59	16,149	**0,877**	0,894**
SD1	Responsibility vs blaming	8	27,49	4,623	0,710	28,15	4,819	0,752	0,721**
SD2	Purposefulness vs goal-undirectted	6	23,34	3,646	0,672	22,35	3,642	0,707	0,867**
SD3	Resourcefulness vs apathy	5	17,86	2,415	0,398	17,97	2,007	0,179	0,395**
SD4	Self-acceptance vs self-striving	10	30,77	7,515	0,867	30,82	5,407	0,650	0,743**
SD5	Congruent second nature	11	40,60	5,867	0,801	40,12	6,480	0,847	0,794**
CO	**Cooperativeness**	36	132,09	11,584	**0,813**	128,94	11,987	**0,817**	0,775**
CO1	Social acceptance vs intolerance	8	29,69	3,132	0,513	29,03	3,503	0,618	0,762**
CO2	Empathy vs social disinterest	5	18,60	2,329	0,284	18,00	2,174	0,299	0,655**
CO3	Helpfulness vs unhelpfulness	8	28,09	2,934	0,300	26,85	3,164	0,364	0,457**

Continued

Table 3.2 Continued

Scales	Character scales	No. of items	Mean	± Standard devaition	**Cronbach's α**	Mean	± Standard devaition	**Cronbach's α**	Test – retest correlation coefficient
CO4	Compassion vs revengefulness	7	26,37	4,747	0,851	26,56	5,052	0,872	0,832**
CO5	Pure-hearted vs self-serving	8	29,34	3,638	0,476	28,50	3,126	0,350	0,600**
ST	**Self-transcendence**	26	79,60	11,617	**0,822**	77,91	11,820	**0,831**	0,630**
ST1	Self-forgetful vs self-conscious	10	30,54	5,923	0,728	30,24	5,990	0,775	0,701**
ST2	Transpersonal identification	8	23,66	4,193	0,700	23,82	3,912	0,616	0,709**
ST3	Spiritual acceptance vs materialism	8	25,40	4,614	0,615	23,85	5,040	0,711	0,626**

Scales	IMPC scales	No. of items	Mean	± Standard devaition	**Cronbach's α**	Mean	± Standard devaition	**Cronbach's α**	Test – retest correlation coefficient
	Psychological climate	**40**			**0.958**			**0.960**	
	Autonomy	**5**	14,88	3,23	0.877	14,32	3,57	0.928	0,804
	Cohesion	**5**	16,29	4,54	0.928	16,82	4,67	0.939	0,704
	Trust	**5**	16,79	5,51	0.949	17,50	5,12	0.924	0,909
	Pressure	**5**	14,88	2,97	0.373	15,24	3,28	0.526	0,604
	Support	**5**	17,47	5,69	0.959	17,85	5,09	0.943	0,890

Recognition	**5**	15,76	3,89	0.697	16,50	3,93	0.738	0,845
Fairness	**5**	16,85	5,17	0.908	17,65	4,83	0.906	0,859
Innovation	**5**	17,32	5,51	0.964	17,88	5,13	0.945	0,858

MBI скали	No. of items	Mean	± Standard devaition	Cronbach's α	Mean	± Standard devaition	Cronbach's α	Test – retest correlation coefficient P<0.001
	22							0.743
Emotional exhaustion	9	21,78	11,80	0.948	22,00	11,46	0.931	0.703
Depersonalization	4	6,69	4,11	0.749	6,88	4,53	0.803	0.791
Personal accomplishment	8	33,34	5,40	0.639	32,91	5,45	0.630	0.826

***Correlation is significant at the 0.001 level (two-tailed).

**Correlation is significant at the 0.01 level (two-tailed).

* Correlation is significant at the 0.01 level (two-tailed).

The most significant issues with such items occur in HA, RD, SD and CO subscales of the TCI-R section, while in IMPC and MBI there were singular items in Pressure, Recognition and PA.

The acceptable levels of internal consistency in the pilot results were found to be lower in the larger sample for all TCI-R subscales (especially for the NS, RD, CO), while in most IMPC subscales (autonomy, trust, pressure, support) as well as in PA of MBI there was increase in Cronbach α values (Table 3.2).

The overall results revealed medium item-item correlation in character and temperament subscales although they also indicated issues with TCI-R subscales (especially in HA, PS, SD, ST) while there was high level of correlation in IMPC and MBI (except for Recognition and PA) (Table 3.3).

The corrected item-total correlation coefficient values provide important information on the quality of a tool. Overall 16 items from the compiled tool require further elaborating since they were found to have a correlation coefficient <0.25 while contributing to increase of more than 0.10 in Cronbach α if deleted.

Further factor analyses are to be performed with a sample size satisfying the requirements (3 to 5 participants per item).

Although the field study results obtained from 302 participants were less satisfying than the pilot run the results were interpreted as convincing of the overall tool reliability and the low correlation and α values in some domains found in our, as well as other, studies are further discussed further on.

A number of the TCI-R temperament and character dimensions in our study were interrelated as expected for functionally related behavior traits (Tables 3.3–3.5). The correlation results are consistent with those in other studies [5, 7–14, 16, 22–25], indicating medium and strong correlations.

The overall and internal consistency of the TCI-R scales was acceptable, except for a few earlier mentioned subscales (CO5 pure-hearted, CO2 empathy, NS3 extravagance, NS4 disorderliness, and RD2 communication). However, similar results have been obtained in various cross-cultural studies in Belgium, France and Serbia, as opposed to findings in the USA suggesting that features inherent in these scales account for the differences rather than cultural factors [5, 7–14, 16, 22–25]. The TCI-R temperament and character scores in the Bulgarian sample were generally comparable with those obtained in Serbia and, to some extent, to those in the USA with some exceptions.

The observed differences may be determined by differences in character traits, which are expected to vary with culture, but are unusual with respect to temperament given the stability of temperament traits across cultures [15, 16].

Table 3.3 Battery internal consistency (inter-scale correlations, Cronbach α) in the field study in Bulgaria (n=302)

Scale	TCI scales	No. of items	Mean	± Standard deviation	Cronbach's α	Item-item correlations (min-max)	Corrected item-total correlations (min-max)
	Temperament scales	240			0.909		
NS	**Novelty-seeking**	**35**	**106.70**	**12.32**	**0.697**	**–0.588 – 0.680**	**–0.248 – 0.495**
NS1	Exploratory excitability vs rigidity	10	31.40	4.46	0.361	–0.322 – 0.461	–0.009 – 0.312
NS2	Impulsiveness vs reflection	9	29.56	5.69	0.678	–0.168 – 0.644	–0.034 – 0.658
NS3	Extravagance vs reserve	9	25.32	4.20	0.288	–0.298 – 0. 474	–0.046 – 0.218
NS4	Disorderliness vs regimentation	7	20.39	3.62	0.298	–0.130 – 0.394	0.056 – 0.197
HA	Harm-avoidance	33	96.71	13.04	0.762	–0.231 – 0.612	0.046 – 0.483
HA1	Anticipatory worry vs optimism	11	32.86	5.37	0.556	–0.184 – 0.608	0.011 – 0.508
HA2	Fear of uncertainty vs confidence	7	20.71	4.20	0.489	–0.116 – 0.523	0.013 – 0.386
HA3	Shyness vs gregariousness	7	20.24	3.80	0.400	–0.193 – 0.528	0.042 – 0.280
HA4	Fatigability and asthenia vs vigor	8	22.93	4.70	0.601	–0.094 – 0.5595	0.156 – 0.420

Continued

Table 3.3 Continued

Scale	TCI scales	No. of items	Mean	± Standard deviation	Cronbach's α	Item-item correlations (min-max)	Corrected item-total correlations (min-max)
RD	**Reward-dependence**	**30**	**99.10**	**10.43**	**0.620**	**–0.309 – 0.543**	**–0.018 – 0.387**
RD1	Sentimentality vs insensitivity	8	28.43	4.42	0.288	–0.078 – 0.272	–0.047 – 0.211
RD2	Openness to warm communication vs aloofness	10	33.49	4.73	0.530	–0.237 – 0.368	–0.050 – 0.361
RD3	Attachment vs detachment	6	18.53	3.54	0.367	–0.154 – 0.380	–0.018 – 0.348
RD4	Dependence vs independence	6	18.64	3.42	0.320	–0.263 - 0.380	–0.131 – 0.279
PS	**Persistence**	**35**	**121.66**	**15.88**	**0.846**	**–0.226 – 0.569**	**0.029 – 0.592**
PS1	Eagerness of effort vs laziness	9	31.77	4.60	0.611	–0.178 - 0.472	–0.100 – 0.531
PS2	Work hardened vs spoiled	8	28.23	4.75	0.693	–0.089 - 0.491	–0.032 – 0.512
PS3	Ambitious vs underachieving	10	34.28	6.15	0.602	–0.043 – 0.385	0.063 – 0.452
PS4	Perfectionism vs pragmatism	8	27.38	4.00	0.503	–0.106 – 0.428	0.046 – 0.377

Scale	Character scales	No. of items	Mean	± Standard deviation	Cronbach's α	Item-item correlations (min-max)	Corrected item-total correlations (min-max)
SD	**Selfdirectedness**	**40**	**124.34**	**17.57**	**0.754**	**–0.224 – 0.613**	**–0.005 – 0.609**
SD1	Responsibility vs blaming	8	23.85	4.99	0.613	–0.052 – 0.453	0.017 – 0.498
SD2	Purposefulness vs goal-undirectted	6	19.80	3.53	0.394	–0.161 – 0.613	–0.024 – 0.341
SD3	Resourcefulness vs apathy	5	14.47	3.62	0.559	–0.060 – 0.523	–0.024 – 0.527
SD4	Self-acceptance vs self-striving	10	30.78	5.93	0.658	–0.139 – 0.431	0.013 – 0.502
SD5	Congruent second nature	11	35.56	6.11	0.651	–0.156 – 0.540	0.084 – 0.516
CO	**Cooperativeness**	**36**	**117.93**	**12.67**	**0.656**	**–0.467 – 0.774**	**–0.294 – 0.586**
CO1	Social acceptance vs intolerance	8	27.03	4.04	0.329	–0.131 – 0.498	–0.003 – 0.300

Continued

Table 3.3 Continued

Scale	Character scales	No. of items	Mean	± Standard deviation	Cronbach's α	Item-item correlations (min-max)	Corrected item-total correlations (min-max)
CO2	Empathy vs social disinterest	5	17.50	2.61	0.242	–0.094 – 0.287	–0.071 – 0.245
CO3	Helpfulness vs unhelpfulness	8	28.29	3.58	0.330	–0.189 – 0.337	–0.117 – 0.270
CO4	Compassion vs revengefulness	7	19.88	5.30	0.707	–0.172 – 0.777	–0.101 – 0.757
CO5	Pure-hearted vs self-serving	8	25.40	3.77	0.115	–0.408 – 0.481	–0.284 – 0.176
ST	**Self-transcendence**	**26**	**80.73**	**12.73**	**0.713**	**0.144 – 0.565**	**0.023 – 0.590**
ST1	Self-forgetful vs self-conscious	10	31.04	5.68	0.667	–0.144 – 0.445	0.184 – 0.443
ST2	Transpersonal identification	8	24.85	4.99	0.689	0.026 – 0.432	0.243 – 0.501
ST3	Spiritual acceptance vs materialism	8	24.89	5.33	0.546	–0.076 – 0.565	0.029 – 0.468

Scales	IMPC scales	No. of items	Mean	±Standard deviation	Cronbach's α	Item-item correlations (min-max)	Corrected item-total correlations (min-max)
	Psychological climate	40			0.942		
A	Autonomy	5	19.05	3.96	0.994	0.973 - 0.995	0.985 – 0.997
C	Cohesion	5	17.83	4.38	0.807	0.885 - 0.991	0.915 – 0.988
T	Trust	5	19.89	4.58	0.984	0.912 – 0.969	0.947 – 0.966
P	Pressure	5	15.91	3.54	0.929	0.515 – 0.974	0.562 – 0.969
S	Support	5	19.48	4.70	0.977	0.940 – 0.972	0.962 – 0.986
R	Recognition	5	17.30	3.46	0.592	–0.158 – 0.622	–0.173 – 0.603
F	Fairness	5	19.18	4.37	0.890	0.482 – 0.757	0.618 – 0.801
I	Innovation	5	19.03	4.57	0.928	0.681 – 0.765	0.795 – 0.846
		22			0.806		

	IMPC	No. of items	Mean	±Standard deviation	Cronbach's α	Item-item correlations (min-max)	Corrected item-total correlations (min-max)
EE	Emotional exhaustion	9	22.06	11.56	0.916	0.340 - 0.770	0.516 – 0.812
DP	Depersonalization	4	6.82	4.55	0.735	0.274 – 0.597	0.438 – 0.633
PA	Personal accomplishment	8	31.87	5.87	0.653	–0.082 – 0.540	0.020 – 0.537

The observed higher NS score in our study exceeding that in Serbia and USA subjects may be reflective of dramatic social and economic changes over the last 20 years, where totalitarian regimes have been replaced by liberalization and entrepreneurship, fostering personal initiative and changes in national cultures and value systems.

SD is associated with psychological, as well as physical wellbeing, expressed as subjective perception and satisfaction. CO is assessed through social tolerance, empathy and willingness to help as well as through the strength of social relations and support that in turn influence subjective well-being and happiness. ST is defined as a potent factor influencing positive emotions and measured by the personal identification with intrinsic values. This functional aspect of personality is most vulnerable and easily affected by hostile and unfavourable environments, and is considered a major factor in burnout development in highly differentiated professions in health care [1,15], as reflected by the higher values in our survey.

Significant differences may be found in higher HA and ST, while RD, SD, CO and PS tend to be much lower than the results obtained in general populations in Serbia and the USA, with USA subjects scoring consistently

Table 3.4 TCI-R scale significant correlations (N=302)

	HA	RD	PS	SD	CO	ST
NS	ns	**0.44****	**0.30****	ns	ns	**0.36****
HA	ns	0.23**	0.12*	ns	ns	**0.29****
RD	ns	ns	**0.48****	0.26**	**0,48****	**0,49****
PS	ns	ns	ns	**0.45****	**0.51****	**0.50****
SD	ns	ns	ns	ns	**0.62****	**0.30****
CO	ns	ns	ns	ns	ns	**0,38****

**. Correlation is significant at the 0.01 level (2-tailed).

*. Correlation is significant at the 0.05 level (2-tailed).

Table 3.5 IMPC scale significant correlations (N=302)

	Cohesion	Trust	Pressure	Support	Recognition	Fairness	Innovation
Autonomy	0.16*	0.19**	0.16*	0.24**	0.21**	0.20**	0.23**
Cohesion	ns	0.47**	0.19**	0.44**	0.29**	0.44**	0.33**
Trust	ns	ns	0.14*	**0.86****	**0.61****	**0.80****	**0.67****
Pressure	ns	ns	ns	0.14*	0.22*		0.12
Support	ns	ns	ns	ns	**0.72****	**0.84****	**0.75****
Recognition	ns	ns	ns	ns	ns	**0.70****	**0.70****
Fairness	ns	ns	ns	ns	ns	ns	**0.77****

**. Correlation is significant at the 0.01 level (2-tailed).

*. Correlation is significant at the 0.05 level (2-tailed).

higher than subjects from other societies in PS, likely reflecting social or biological factors specific to the USA [4, 5, 22–24].

Our specific results with observed lower SD and CO yet higher ST scores, point to a speculative but plausible interpretation related to high stress and burnout syndrome levels in the studied medical professions.

The observed levels may reflect the similarities in cultural models of the two neighboring countries. The so-called 'Balkan temperament' has its specific geographical and historical determinants. Because of the common historical background, including a long period of foreign political dominance, the people in these regions tend to be more pragmatic and conservative perceiving reality as a survival challenge rather than an opportunity for creativity or novelty. New experience is often referred to with suspicion or pessimism. Individuals often experience an inner conflict between CO and SD, where CO is considered a luxury and a final resort. Thus, personality faces the challenge of emotionality or frustration. There is a pragmatic subconscious drive to interpret selfishness as a life strategy yet not useful in personal development and self-preservation.

A number of the TCI-R temperament and character traits were interrelated. For instance, PS was correlated with CO and ST that correlated with RD.

Although these interrelated traits reflect different personality processes, some more emotional and closer to their biological roots, for example HA and PS, and some more rational and adaptive in nature, for example ST and CO, they interact to shape expressed behaviours.

The national social psychological profile is formed by a complex composition of social, historical and cultural factors in a certain geographical region. Bulgaria has always been defined as 'cross-roads' of different civilizations imprinting their typical features on ethnicity, religiousness, everyday practices and culture [26]. The internalized regional sociocultural perceptions are reflected by the individual and the overall group answers to TCI-R.

In general, the cross-cultural adaptation process, even if carried out in a rigorous way, does not always lead to the best target version and suggests that it would be useful to develop new scales specific to each culture and, at the same time, to think about transcultural adaptation.

During this cross-cultural adaptation process, items were not modified to adapt the original questionnaire to Bulgarian culture but rather revised to follow closely the source version. Some questions may not be applicable as formulated to all of the Bulgarian population, such as the ST subscale question referring to spirituality, SD, HA and RD scales were most of the non-consistent items were found. On the other hand, some questions involving the same items posed differently in different dimensions gave rise to confusion or the impression of repetition in the Bulgarian version.

However, the overall internal consistency of the battery and its scales as well as the interscale and test–retest correlations provide response to accept that the translated version of the questionnaire is acceptable and cross-culturally applicable further research with the purposes of studying organizational stress and burnout risk in health-care professionals.

The field study results provide grounds for this assumption supporting the working hypothesis and conceptual model exploring relation between unfavorable psychological climate, manifestations of burn out and the personality profile of the employees [19, 20, 26] (Table 3.6).

Table 3.6 MBI scale significant correlations (N=302)

	DP	PA
EE	**0.54****	**-0.22****
DP	ns	-0.18**

**. Correlation is significant at the 0.01 level (2-tailed).
*. Correlation is significant at the 0.05 level (2-tailed).

Table 3.7 Correlations between burnout components and study variables

Scales	Emotional exhaustion	Depersonalization	Personal accomplishment
IMPC			
Autonomy	ns	ns	**0,36****
Cohesion	ns	–0,15*	0,20**
Trust	–0,17**	–0,20**	0,22**
Pressure	**0,46****	0,27**	ns
Support	–0,22**	–0,18**	0,26**
Recognition	ns	ns	0,14*
Fairness	–0,18**	ns	0,21**
Innovation	–0,15*	ns	0,22**
TCI-R	ns	ns	ns
NS	ns	ns	ns
HA	0.27**	0.14*	ns
RD	ns	ns	0,17**
PS	ns	ns	**0,37****
SD	ns	ns	ns
CO	ns	ns	0,17**
ST	0.17**	ns	ns

**p < .01; *p < .05.

The established correlations present evidence on the nature of the tested variables and determinants. They reveal the relation of professional Autonomy to Personal accomplishment that both essential in medical professions.

Professional exhaustion may be interpreted as impaired balance between environmental demands and the available resources of the individual The Pressure in organizational context of large health-care establishments may expectedly affects Emotional exhaustion of personnel, while Support, Fairness and Innovation may prevent it as well as contribute to the perception of higher Personal accomplishment as do Cohesion, Trust and Recognition.

Some personality traits prove to be positively related to burnout dimensions and may be acting as mediators of organizational climate effects. We may speculate that HA and ST are both positively related to Emotional exhaustion in reference to the high risk and demanding nature of medical professions, while RD, PS and CO relate to Personal accomplishment associated with the significance of health outcomes, team work and patient-doctor relations. These features in reference to group characteristics are subject to discussion in the respective chapters.

References

[1] Sperber, A. D. (2004) Outcome assessment. Translation and validation of study instruments for cross-cultural research. Available at: http://www.sciencedirect.com/science/article/pii/S0016508503015646#bib1001 (last accessed 31 May 2012).

[2] Endacott, R., Benbenishty, J. & Seha, M. (2010) Preparing research instruments for use with different cultures. Intensive and Critical Care Nursing, 26, 64–68.

[3] Mapi Research Institute (2002) Linguistic validation of the PedsQL – a quality of life questionnaire research and evaluation, limited use translation of PedsQ. Available at: http://www.pedsql.org/PedsQLLinguistic-Validation-Guidelines.doc (last accessed 31 May 2012).

[4] Cloninger, C. R., Przybeck, T. R., Svrakic, D. M. & Wetzel, R. D. (1994) The Temperament and Character Inventory – A Guide to Its Development and Use. St. Louis: Washington University.

[5] Cloninger, C. R., Svrakic, D. M. & Przybeck, T. R. (1993) A psychobiological model of temperament and character. Archives of General Psychiatry, 50, 975–990.

[6] Snopek, M., Snopeka, M., Hublovaa, V., Porubanovaa, M. & Blatny, M. (2012) Psychometric properties of the temperament and character

inventory-revised (TCI-R) in Czech adolescent sample. Comprehensive Psychiatry, 53, 71–80.

[7] Duijsens, I. P., Spinhoven, P., Goekoop, J. G., Spermon, T. & Eurelings-Bontekoe, E. H. M. (2000) The Dutch temperament and character inventory (TCI): dimensional structure, reliability and validity in a normal and psychiatric outpatient sample. Personality and Individual Differences, 28, 487–499.

[8] Otter, C., Huber, J. & Bonner, A. (1995) Cloninger's tridimensional personality questionnaire: reliability in an English sample. Personality and Individual Differences, 18 (4), 471–480.

[9] Pelissolo, A. & Lepine, J. P. (2000) Normative data and factor structure of the temperament and character inventory (TCI) in the French version. Psychiatry Research, 94, 67–76.

[10] Sung, S. M., Kim, J. H., Yang, E., Abrams, K. Y. & Lyoo, I. K. (2002) Reliability and validity of the Korean version of the temperament and character inventory. Comprehensive Psychiatry, 43, 235–243.

[11] Zakrzewska, M., Samochowiec, J., Rybakowski, F., Hauser, J. & Pelka-Wysiecka, J. (1994) Polish version of temperament and character Inventory (TCI): the analysis of reliability. Psychiatria Polska, 35, 455–465.

[12] Arkar, H., Sorias, O., Tunca, Z., Safak, C., Alkin, T., Binnur, A. B., Sahin, S., Akvardar, Y., Sari, O., Ozerdem, A. & Cimilli, C. (2005) Factorial structure, validity, and reliability of the Turkish temperament and character inventory (in Turkish). Turk Psikiyatri Dergisi, 16, 190–204.

[13] Brändström, S., Schlette, P., Przybeck, T., Lundberg, M., Forsgren, T., Sigvardsson, S., Nylander, P., Nilsson, L., Cloninger, R. & Adolfsson R. (1998) Swedish normative data on personality using the temperament and character inventory. Comprehensive Psychiatry, 39, 122–128.

[14] Brändström, S., Sigvardsson, S., Nylander, P. & Richter, J. (2008) The Swedish version of the temperament and character inventory (TCI): a cross-validation of age and gender influences. European Journal of Psychological Assessment, 1 (24), 14–21.

[15] Kalinov, K., Milanova, V. & Jablenski, A. (2005) Validity and factor structure of temperament and character questionnaire in Bulgarian population. (in Bulgarian). Psihologicheski Izsledvania, 2, 57–69.

[16] Dzamonja-Ignjatovic, T., Svrakic, D., Svrakic, N., Jovanovic, M. & Cloninger, R. (2010) Cross-cultural validation of the revised temperament and character inventory: Serbian data. Comprehensive Psychiatry, 50, 649–655.

[17] Pelissolo, A., Ecochard, P. & Falissard, B. (2008) Psychometric characteristics of Cloninger's criteria for personality disorder in a population of French prisoners. International Journal of Methods in Psychiatric Research, 17 (1), 30–34.

[18] Koys, D.J. & DeCotiis, T.A. (2001). Inductive measures of psychological climate. Human Relations 44, 265–285.

[19] Maslach, C., Jackson, S.C. & Leiter, M.P. (1996). Maslach Burnout Inventory. Palo Alto, California: Consulting Psychologists Press.

[20] Maslach, C., Schaufeli, W.B. & Leiter, M.P. (2001). Job burnout. Annual Review of Psychology 52, 397–422.

[21] Norman, G. (2010). Likert scales, levels of measurement and the "laws"of statistics. Advances in Health Sciences Education, 15(5), 625–632. doi:10.1007/s10459-010-9222-y

[22] Cloninger, C. R., Przybeck, T. R., Svrakic, D. M. & Wetzel, R. D. (1999) The Temperament and Character Inventory – Revised. St. Louis: Washington University.

[23] Miettunen, J., Kantojärvi, L., Veijola, J., Järvelin, M. & Joukamaa, M. (2006) International comparison of Cloninger's temperament dimensions. Personality and Individual Differences, 41, 1515–1526.

[24] Hansenne, M., Delhez, M. & Cloninger, C. R. (2005) Psychometric properties of the temperament and character inventory – revised (TCI-R) in a Belgian sample. Journal of Personality Assessment, 85 (1), 40–49.

[25] Naito, M., Kijima, N. & Kitamura, T. (2000) Temperament and character inventory (TCI) as predictors of depression among Japanese college students. The Journal of Clinical Psychiatry, 56 (12), 1579–1585.

[26] Stoyanov, D., Stoikova, M., Tornyova, B., Tilov, B. G., Turnovska, T., Mateva, N. & Hyusein, N. (2011) Burn out syndrome in health care personnel: ethical, psychological and methodological implications. 5th Balkan Congress on the History and Ethics of Medicine 11-15. X. 2011, Istanbul, pp. 449–462. ISBN 978-975-420-871-9.

4

Comparative Study of Burnout in Medical Professionals from Psychiatric Units and Other Health Care Sectors

Boris Tilov PhD, Assistant Professor of General Psychology
Department of Health Care Management,
Faculty of Public Health, MUP

Biyanka Tornyova PhD, Associate Professor of Medical Education
Department of Health Care Management,
Faculty of Public Health, MUP

Mariya Semerdzhieva MD, PhD, Associate Professor of Medical Management and Medical Etiquette, Department of Health Care Management, Faculty of Public Health, MUP

Drozdstoy Stoyanov MD, PhD, Full Professor of Psychiatry, Medical Psychology and Person Centered Medicine,
Faculty of Medicine, MUP, Vice Chair Executive Committee PSIG,
Royal College of Psychiatrists, Visiting Fellow, University of Pittsburgh

Work is one of the main sources of social satisfaction, but it can also cause the occurrence of stress [1,2]. Work-related stress in itself does not lead to exhaustion. However, those who are in a highly stressful work environment, such as medical staff in intensive care units, may experience higher levels of anxiety, anger, behavioral disturbances and depression symptoms [3]. Empirical studies demonstrate that the negative affectivity is a moderator of emotional dissonance and emotional exhaustion. Several studies have shown that physicians are in risk of developing high levels of stress, experience less job satisfaction and have threatened mental health [4, 5, 6].

In cases when autonomy is removed, high demands, accuracy, workload and responsibility for other individuals may become an unbearable burden. A study of the relationship between the level of burnout, depression and

Drozdstoy St. Stoyanov (Ed.), New Model of Burn Out Syndrome: Towards Early Diagnosis and Prevention, 47–58.

satisfaction with life and work of physicians in emergency units in Canada shows that with age, the organization of free time outside clinical practice influences job satisfaction and emotional well-being [7].

Garrett and McDaniel, claim that the perception of uncertainty in the environment predicts burnout. The social climate in the workplace is decreasingly associated with burnout, but support at the workplace can prevent the onset of the syndrome [8].

In their study, Bellani et al. found correlations between "burnout", "personal achievement", anxiety and depression. According to them, anxiety, depression, emotional reactions, attitudes, expressions of ego and the ability for interpersonal relationships and teamwork are important factors that determine the profile of a strong "burnout" and "reduced personal achievements" in healthcare professionals [9].

McKnight and Glass noted a significant increase in burnout syndrome and depressive symptoms in their two-year study of 100 nurses. The deviation shared between the burnout and depression (20%) may be due to their co-development. According to the initial assessment and subsequent results, the nurses with burnout symptoms demonstrate accurate perception of lack of work control, while nurses who are not affected by burnout, overestimate the work control. The accuracy of perception increases in a linear progression with the degree of burnout, regardless of the depression symptoms [10].

A study in China shows that more and more physicians quit job or intend to change their workplace due to dissatisfaction with the provided work conditions. Zhang and Feng have studied the relationship between psychological climate, burnout syndrome and the intention for change (turnover). The results established that of 1 600 surveyed physicians, the turnover was in downward strong correlation for all job satisfaction subscales, and upward for each of the burnout syndrome subscales. Psychological climate, satisfaction with remuneration and rewards, satisfaction with the organization and management, and emotional exhaustion were identified as significant direct predictors of turnover in physicians. This suggests that the change in psychological climate may lead to decrease of physicians' turnover intentions due to burnout [11].

In a study of Ogresta, Rusac and Zorec, regarding the importance of satisfaction and manifestations of stress at the workplace in healthcare employees in psychiatry ward, found that they are exposed to moderate degree burnout syndrome. Statistically significant difference for their occupation was not established. Overall, both, manifestations of job satisfaction and manifestations of stress at the workplace, proved to be predictors of burnout syndrome. The gradual regression analysis showed that satisfaction with remuneration and rewards, psychological climate, development opportunities, degree of workplace stress psychological and physical manifestation are significant predictors of emotional exhaustion. Satisfaction with the psychological climate is a significant predictor of lower levels of personal achievements [12].

4.1 Material and Methods

In this section we will examine and compare two groups of healthcare professionals. The first group, or so called "General Group", consists of employees in several fields: oncology, palliative care, dentistry, toxicology, orthopedics, and rehabilitation.

The second group consists of healthcare professionals occupied in psychiatric wards and units. The consistency of the first group is based on the insufficient number of persons in the different fields of healthcare, while the size of the second group is considered sufficient for the objectives of this study.

In the distribution of average values between the two groups we noticed that those working in the psychiatric wards have higher values for "Self-transcendence" (u = 2.36; P <0,05) and "Reward dependence" item (u = 2.47; P <0, 05) (Table 4.1).

Table 4.1 Comparison of the Surveyed Groups by Average Values Obtained through Temperament and Character Inventory (Revised Version)

TCI-R	Groups	N	N items	MD ± SE	SD	u	P
Novelty seeking	Others	77	35	106.08 ± 1.26	11.02	1.06	>0.05
	Psychiatrists	25	35	108.48 ± 1.88	9.40		
Harm avoidance	Others	77	33	95.27 ± 1.28	11.27	1.72	>0.05
	Psychiatrists	25	33	89.56 ± 3.07	12.05		
Reward dependence	Others	77	30	99.01 ± 1.15	10.05	2.47	<0.05
	Psychiatrists	25	30	93.04 ± 2.12	15.35		
Persistence	Others	77	35	119.26 ± 1.76	15.41		
	Psychiatrists	25	35	114.72 ± 3.73	10.60	1.10	>0.05
Self-directedeness	Others	77	40	122.09 ± 1.85	16.25	1.00	>0.05
	Psychiatrists	25	40	126.96 ± 4.48	22.40		
Cooperativeness	Others	77	36	117.65 ± 1.45	12.75	1.22	>0.05
	Psychiatrists	25	36	114.16 ± 2.46	12.30		
Self-transcendence	Others	77	26	81.50 ± 1.32	11.51	2.36	<0.05
	Psychiatrists	25	26	71.68 ± 3.94	12.00		

Table 4.2 Comparison of the Surveyed Groups by Average Values Obtained by Inductive Measurement of Psychological Climate

IMPC	Groups	N	N items	MD±SE	SD	u	P
Autonomy	Others	18	5	18.20 ± 0.77	3.84	0.11	>0.05
	Psychiatrists	36	5	18.10 ± 0.47	4.14		
Cohesion	Others	18	5	15.44 ± 0.87	4.35	2.56	<0.05
	Psychiatrists	36	5	18.03 ± 0.52	4.60		
Trust	Others	18	5	18.16 ± 0.89	4.43	2.61	>0.05
	Psychiatrists	36	5	20.79 ± 0.47	4.10		
Pressure	Others	18	5	15.64 ± 0.63	3.15	0.84	>0.05
	Psychiatrists	36	5	15.03 ± 0.36	3.13		
Support	Others	18	5	17.52 ± 0.92	4.61	2.92	>0.05
	Psychiatrists	36	5	20.56 ± 0.44	3.91		
Recognition	Others	18	5	15.56 ± 0.56	2.87	2.70	>0.05
	Psychiatrists	36	5	17.35 ± 0.36	3.14		
Fairness	Others	18	5	16.84 ± 0.84	4.22	3.35	>0.05
	Psychiatrists	36	5	20.05 ± 0.46	4.01		
Innovation	Others	18	5	17.00 ± 0.84	4.19	2.51	>0.05
	Psychiatrists	36	5	19.42 ± 0.47	4.13		

When we compared the results we established higher averages and a statistically significant difference in the psychiatrists group for "Solidarity" (u = 2.56; P <0.05) (Table 4.2).

In this comparison we did not establish statistically significant differences between the two groups.

Employees of psychiatric wards and units are more emotionally exhausted, depersonalized and with lower levels of performance.

When we analyzed the results of separate scales, by considering their clarity and effect, we established that in the first group of 42.9% of the respondents have low values of emotional exhaustion, 29.9% of the same group have moderate values, and 27.3% suffer high mental and emotional exhaustion.

The values indications for the second group of respondents, those working in psychiatric wards, indicate that 40% of respondents suffer from high levels of emotional exhaustion, and 36% of them have low emotional exhaustion levels.

When we compared levels of emotional exhaustion of healthcare professionals at psychiatric wards and units, we noted higher indicators than in oncology, palliative care and dentistry healthcare professionals.

Table 4.3 Comparison of the Surveyed Groups by Average Values Obtained by Maslach Burnout Inventory for Burnout Syndrome Diagnostics

MBI	Groups	N	N items	MD ± SE	SD	u	P
Emotional exhaustion	Others	36	9	19.31 ± 1.18	10.38	1.11	>0.05
	Psychiatrists	18	9	22.40 ± 2.53	12.62		
Depersonalization	Others	36	4	6.75 ± 0.51	4.46	1.09	>0.05
	Psychiatrists	18	4	7.88 ± 0.90	4.50		
Personal accomplishement	Others	36	8	31.27 ± 0.72	6.30	0.39	>0.05
	Psychiatrists	18	8	30.76 ± 1.08	5.39		

In the "Depersonalization" scale we established inverse correlation in the first group, where the low levels percentage is 50.6% of the surveyed group, 40.3% fall in the moderate level of the scale, and only 9.1% show high levels of this measure (Table 4.3).

The values in the second surveyed group are as follows: 48% of psychiatric wards professionals fall within the moderate levels of depersonalization, 40% have low levels, and 12% have high levels of this measure.

From the obtained results we established higher moderate levels in the first group. Higher levels of "Depersonalization" item are shown by the group of psychiatrists from the first group.

The analysis of the third scale, "Performance", in the first group shows that the largest percentage is in the high levels with 51.9% compared with 10.4% for lower levels, and 37.7% for average levels.

The results of the second group, professionals occupied in psychiatric wards, indicated highest percentage of performance with 56% of respondents in the high level, 36 % in moderate levels, and 8% in lower levels.

In summary, we can say that professionals occupied in psychiatric wards are more emotionally exhausted and depersonalized, but have higher performance comparing with professionals in oncology, palliative care, dentistry, rehabilitation and other fields.

In the results of Table 4.4 we find high levels of inverse correlation between MBI scale "Emotional Exhaustion" and the IMPC scales "Trust" ($r = -0.587; P < 0.001$), "Recognition" ($r = -0.510; P < 0.001$) and "Solidarity" ($r = -0.453$; $P < 0.001$) scales. These results cannot show us what the cause and the effect are, but illustrate that the more decreased interdependence, trust, lack of staff and recognition for work done by healthcare teams working in oncology, toxicology, palliative care, etc. are, the more increased emotional exhaustion is. We have established inverse correlation in the first group of emotional exhaustion for all IMPC scales, as for the "Support" (r = –0.345; P < 0.01), "Honesty" (r = –0.344; P < 0.01), and "Innovation" (r = –0.344; P < 0.01) scales we established moderate inverse correlation. When healthcare professionals have supportive environment at their workplace, their work is fairly recognized, and innovation and variety of workflow is present, this is a type of projection against emotional exhaustion. The levels for "Autonomy" (r = –0.230; P < 0.05) and "Pressure" (r = 0.308; P < 0.05) scales are low. Therefore we can conclude that the lack of independence and freedom of action, and systemic pressure on workflow can lead to development of higher levels of emotional exhaustion. The more you observe the rules in a structure, a team or group, regardless of the regulatory status, while encouraging creativity, the more limited the development of emotional exhaustion among professionals in oncology, palliative care and dentistry wards is.

In the analysis of "Depersonalization" scale, we can highlight the moderate inverse correlation with "Solidarity" (r = –0.355; P < 0.01), "Support" (r = –0.306; P < 0.05) and "Trust" (r = –0.285; P < 0.05) items. Also, the pressure and the lack of individual persistence feature underlie the development of the non-humane component. In summary, we note that the lack of a supportive environment combined with solidarity and trust presuppose the occurrence of depersonalization.

When comparing the correlation of the third scale – "Performance", we can claim that "Persistence" (r = 0.412; P < 0.001), "Support" (r = 0.325; P < 0.01), "Autonomy" (r = 0.315; P < 0.01), and "Solidarity" are prerequisite

Table 4.4 Correlation between MBI and the TCI-R and IMPC items for the First Surveyed Group

	Emotional exhaustion	Depersonalization	Performance
Reward dependence	–	–	0.252 *
Persistence	–	–.225 *	0.412***
Autonomy	–0.230*	–	0.315**
Cohesion	–0.453 ***	–0.355 **	0.299**
Trust	–0.587 ***	–0.285 *	0.236*
Pressure	0.308 *	0.245 *	–
Support	–0.345 **	–0.306 *	0.325**
Recognition	–0.510 ***	–	–
Fairness	–0.344 **	–	0.290*
Innovation	–0.344 **	–	–

* Correlations with significance level $p < 0.05$.

** Correlations with significance level $p < 0.01$.

*** Correlations with significance level $p < 0.01$.

Table 4.5 Correlation between the MBI Scales and Separate items of TCI-R and IMPC for the Second Group – Psychiatric Wards

	Emotional exhaustion	Depersonalization	Performance
Harm avoidance	–	–	–0.705 ***
Persistence	–	-	0.500*
Self-directedeness	–	–	0.467*
Self-transcendence	–	0.470*	0.605**
Autonomy	–	-	-
Cohesion	–	- 0.637 *	–
Trust	- 0.442*	- 0.551 **	–
Support	- 0.453 *	- 0.551 **	–
Recognition	–	- 0.624 **	–
Fairness	- 0.409 *	- 0.550 **	–
Innovation	- 0.449 *	- 0.674 **	- 0.411*

* Correlations with significance level $p < 0.05$.

** Correlations with significance level $p < 0.01$.

*** Correlations with significance level $p < 0.01$.

for increase of performance rate of employees working in hazardous healthcare environments such as oncology, palliative care, toxicology, etc.

Compliance and enforcement of the rules established for all employees and display of tolerance towards errors is another major factor for performance increase. Accordingly, the respondents stated that their work activity and performance are influenced by both fair treatment and adequate material compensation for their efforts.

In the analysis of the second surveyed group of healthcare professionals occupied in psychiatric wards, we established weaker dependence, comparing to the first group, between separate TCI-R and IMPC scales concerning emotional exhaustion scale. What we established as a significant difference between both groups is that in the second group, there is high inverse correlation dependence by group features with "Depersonalization" scale and moderate to high correlation with "Performance" scale.

For the TCI-R questionnaire, we noted the following scales measuring character dispositions with high correlation with "Performance" scale. This is inverse correlation to "Avoid Damage" (r = –0.705; P < 0.001) with performance. The analysis shows that the more individual employees act professionally without fixation and fear of certain hazard and damage, the greater the level of performance. Accordingly, another connection that we found for increasing performance level among employees in psychiatric wards is the personal characteristics for "Self-transcendence" (r = 0.605; P < 0.01),

"Persistence" (r = 0.500; P < 0.05) and "Self-directedeness" (r = 0.467; P < 0.05) (Table 4.5).

Thus, the results indicate that the more an employee of the psychiatric ward or unit is pessimistic, hesitant, anxious and stranded, the lower are the performance levels, and vice versa, the more a person is confident and stable, the more it is prone to high professional results. Accordingly, the "Autonomy" scale shows the employees that have more freedom to operate in concern with rules and regulations at the workplace are more likely to identify themselves as able to perform.

In the analysis of the correlation between "Depersonalization", individual and group features, we found a high inverse correlation with psychological climate. "Solidarity" (r = –0.637; P < 0.05), "Recognition" (r = –0.624; P < 0.01), "Innovation" (r = –0.674; P < 0.01), and "Trust" (r = –0.551; P < 0.01), "Support" (r = –0.551; P < 0.01), and "Justice" (r = –0.550; P < 0.01), combined with individual striving for "Self-transcendence" (r = 0.407; P < 0.01) implies dealing with depersonalization as a whole and non-humane attitude towards patients. The more employees in the psychiatric sector are unprincipled and vindictive, without desire

Table 4.6 ANOVA Analysis of the Effect of Group and Individual Factors on the Professional Exhaustion Level in the First and Second Surveyed Groups

	MBI Scales					
Scales TCI-R, IMPC	Emotional exhaustion		Depersonalization		Performance	
	F	P	F	P	F	P
	First group – oncology, palliative care, dentistry, etc.					
Reward dependence	–	–	–	–	2.90	< 0.01
Self-directedeness	–	–	1.90	< 0.05	–	–
Cohesion	2.18	< 0.05	–	–	1.90	< 0.05
Trust	4.27	< 0.001	–	–	–	–
Fairness	2.63	< 0.01	–	–	–	–
	Second group – psychiatric wards					
Self-directedeness	–	–	–	–	9.55	< 0.01
Autonomy	–	–	–	–	3.40	< 0.05
Trust	–	–	4.19	< 0.01	–	–

for professional development, combined with a lack of understanding and imagination, the more non-humane they become in daily contact with patients.

In the analysis of "Emotional Exhaustion" scale, we established a moderate inverse correlation with group features as "Support" ($r = -0.453$; $P < 0.05$), "Innovation" ($r = -0.449$; $P < 0.05$), "Trust" ($r = -0.442$; $P < 0.05$) and "Justice" ($r = -0.409$; $P < 0.05$). This correlation shows us that the lack of open communication, fear of retaliation, lack of teamwork and innovative approaches, duality in applying rules, and the lack of a creativity positive evaluation are evident predictor for the occurrence of emotional exhaustion in professionals of psychiatric wards.

The factor analysis showed the factorial significance of the individual scales of the TCI-R and IMPC questionnaires and the MBI "Emotional Exhaustion", "Depersonalization", and "Performance" scales. In this analysis we established that for the first group there are four scales affecting the occurrence of professional exhaustion among workers in oncology, palliative care, dentistry, etc. In the ANOVA we found greater dependence of group functioning, as opposed to the individual functioning.

For group psychological climate factors we established correlation with the emotional exhaustion in the first surveyed group, "Solidarity", "Trust" and "Justice". Regarding "Depersonalization" there was a statistically significant correlation between first group individual features such as "Self-directedeness", while for the second group it was "Trust". For the third feature, "Performance", we established correlation with "Solidarity" for first group, and "Self-directedeness" and "Autonomy" for the second group (Table 4.6).

4.2 Discussion

Low levels of health care organizations professionals' adaptation mechanisms in stressful environment sets for a higher anxiety and difficult adaptability, leading in turn to high levels of emotional exhaustion and is a prerequisite for faster professional exhaustion.

Burnout syndrome and disorders associated with stress are common among medical professionals, but it suggests that some healthcare professionals are more prone to BS than others. This study aims to determine the BS intensity among two groups of physicians: psychiatrists and general group of oncologists, toxicologists, dentists, therapists and others.

BS affects personal well-being and job performance. It is important to identify the person prone to its development in order to take preventive measures, such as stress management and improved coping strategies in both surveyed groups.

The data show that the more an individual employee in these fields feels the lack of unity and cohesion, the more likely he or she is to reach emotional exhaustion and vice versa. In cases of interdependence, sense of teamwork, recognition and justice, the level of emotional exhaustion declines. Open communication and expression of tolerance among staff on the one hand and with management on the other, also affects the decrease of emotional exhaustion levels.

References

[1] Pearlin, L.I., Schooler, C., 1978. The structure of coping. J. Health Soc. Behav. 19, 2–21;

[2] Rotter, J.B., 1966. Generalized expectancies for internal versus external control of reinforcement. Psychol. Monogr. Gen. Appl. 80 (1), 1–28.

[3] Chamberlain, K., Zika, S., 1990. The minor events approach to stress: support for the use of daily hassles. Br. J. Psychol. 469–481.

[4] Abraham, R., 1999. Negative affectivity: moderator or confound in emotional dissonance-outcome relationships? J. Psychol. 133 (1), 61–72.

[5] Hsy K, Marshall, V. Prevalence of depression and distress in a large sample of Canadian residents, Interns and Fellows. Am. J. Psychiatry 1987; 144: 1561–1566;

[6] Sutherland V., Cooper CL. Job stress, satisfaction, and mental health among general practitioners before and after introduction of new contract. Br. Med. J. 1992; 13(304 6841): 1545–1548.

[7] Lloyd S, Streiner D, Shannon S. Burnout, depression, life and job satisfaction among Canadian emergency physicians. J. Emerg. Med. 1994: 12(4): 559–565.

[8] Garrett DK, McDaniel AM. A New Look at Nurse Burnout: The Effects of Environmental Uncertainty and Social Climate Journal of Nursing Administration: 2001 Feb; 31(2): 91–96

[9] Bellani ML, Furlani F, Gnecchi M, Pezzotta P, Trotti EM, Bellotti GG. Burnout and related factors among HIV/AIDS health care workers. AIDS Care. 1996 Apr; 8(2): 207–21.

[10] McKnight JD, Glass DC. Perceptions of control, burnout, and depressive symptomatology: a replication and extension. J Consult Clin Psychol. 1995 Jun; 63(3): 490–4.

[11] Zhang Y, Feng X. The relationship between job satisfaction, burnout, and turnover intention among physicians from urban state-owned medical institutions in Hubei, China: a cross-sectional study. BMC Health Serv Res. 2011; 11: 235

[12] Ogresta J, Rusac S, Zorec L. Relation between burnout syndrome and job satisfaction among mental health workers. Croat Med J. 2008 Jun; 48 (3): 364–74

5

Burnout in Healthcare Employees Working in Surgical Departments, Anesthesiology and Intensive Care

Rositsa Dimova MD, PhD, Head Assistant Professor of Social Medicine
Department of Health management, Health Economics
and General Practice,
Faculty of Public Health, Medical University of Plovdiv

Nonka Mateva PhD, Associate Professor of Social Medicine,
Mathematician and Statistical Advisor
Department of Health Management, Health Economics
and General Practice,
Faculty of Public Health, Medical University of Plovdiv

Dessislava Bakova PhD, Senior Assistant Professor of Health Care
Management
Department of Health Care Management,
Faculty of Public Health, Medical University of Plovdiv

Boris Tilov PhD, Assistant Professor of General Psychology
Department of Health Care Management,
Faculty of Public Health, Medical University of Plovdiv

5.1 Introduction

The topic of increased risk from burnout syndrome development in specialists working in intensive care units and surgical departments is especially topical due to the significance of the problem and the necessity of taking urgent relevant measures to limit its development (Lyochkova M. 2004, Balch CM. 2010, Embriaco N. 2007, Petrova G. 2005).

Prevalence rates range from 10% to 50%, depending on profession, assessment tools and population. Generally, both women and men are equally

Drozdstoy St. Stoyanov (Ed.), New Model of Burn Out Syndrome: Towards Early Diagnosis and Prevention, 59–70.

affected by burnout (Brand S, Holsboer-Trashler E. 2010). Experience has shown that among the most susceptible groups of medical professionals (physicians, nurses, laboratory assistants etc.) are mainly those who work in intensive care units, oncological, psychiatric, palliative care and surgical departments, in general medical practice and in emergency medical care (Petrova G. 2003, Cloninger CR. 2011, Embriaco N. 2007, Gundersen L. 2001, Klimo JP. 2012, Ksiazek I. 2011, Lloyd S. 1994, Maffasioli DG. 2010, Maslach C. 1982).

In the last few years, in the specialized literature a number of evidence have been described and widely discussed, they are connected with the factors of the working environment and the approaches for the prevention, origin and development of burnout syndrome in susceptible healthy specialists (Todorova M. 2005, Bohle A. 2001, Känel R.2008, Keller KL. 1989, Schwartz RW. 1994).

The role of the stressogenic factors of the work environment, of the psychosocial climate, and of the temperament-personality features for burn out occurrence, is unquestionable (Petkova M. 2003, Petrova G. 2003, Todorova M. 2005, Shopov D. 2009, Cloninger CR. 2011, Lloyd SD. 1994). Numerous published monographs, dissertations, national, inter- and cross-cultural studies indicate this. Only in the last 4–5 years the list of specialized literature on this problem has increased significantly (Estryn-Behar M. 2010, Garcia-Izquierdo M. 2012, Gundersen L. 2001, Ksiazek SnI. 2011).

It has been proved that the nature of the medical labour and the work conditions place medical specialists on one of the first places in the classification among professions exposed to the highest levels of stress (Bonev I. 2010, Lyochkova M. 2004, Bohle A. 2001, Embriaco N. 2007). Bering in mind the specific factors of the working environment in the accident and emergency units, intensive care units and surgical clinics, there is irrefutable evidence that the organizational psychoclimate is one of the leading factors for the risk of 'emotional burnout' occurrence (Shopov D . 2009, Maslach C. 1982). A number of researchers have worked in this field and they have ascertained and assessed the power of the causative dependence and the direction of the relation between the stressogenic factors of the working environment and the psycho-social changes among medical specialists and even the incidence of psycho-somatic diseases (Embriaco N. 2007).

The specific factors of the working environment are more likely to be perceived as normal rahter than as extreme conditions (Hansen N. 2009, Potter C. 2006). They include: team work, high degree of personal responsibility, promptness and timeliness (not postponement) in the implementation of medical services; great number of patients and extra work load; making important medical decisions in the conditions of high degree of uncertainty and risk in limited and/or missing information; violence and aggression on the side of the patients and /or on the side of their relatives; uncontrollable circumstances and complicated situations; bureaucracy etc. (Maslach C. 1982, Potter C.2006). The duration and structure of duties are another important determinant which influences significantly the subjective well-being and the degree of job satisfaction among the medical staff from the accident and emergency wards (Potter C.2006).

In his article Gundersen L. reports that as a defense mechanism 73% of physicians cited "daily interaction with patients" as the most important or rewarding aspect of practicing medicine (Gundersen L. 2001).

Stressogenic factors frequently cause lack of tolerance; fear of making mistakes; hetero- or auto-aggression; deteriorated communication among medical specialists with patients, as well as with their relatives; low job satisfaction; frustration; neurotic conditions and development of specific diseases (Gundersen L. 2001, Hansen N. 2009, Potter C. 2006, Pejušković B. 2011). The result of this is the described comparatively high degree of dependence to different narcotic substances and/ or suicides among medical professionals, working in the risk departments (Klugel MT. 1999).

The analysis of other Bulgarian authors suggests that the profile of professional burnout in different professions has different configuration (Nenova A. 2005). We agree with this statement and support the thesis that burnout phenomenon relates in highest degree to the medical professions and specialties which are highly specialized and risky with a view of the nature of their work and occasionally the unpredictability of the outcome of the medical aid (Bonev I. 2010, Hansen N. 2009, Klimo JP. 2012, Petrova G.2005).

On the other hand, the lack of a uniform working model, which combines in one the three constructs (work conditions, psycho-climate and character–personality traits of the individual), which lead to burnout occurrence, is the cause for conducting a research such as ours. Additionally, the relation between the separate determinants (internal- personality and external-psycho-social risk factors of the working environment) for occurrence of work-related stress in healthy specialists, in those working in accident and emergency units, intensive care units and surgical clinics (Potter C. 2006, Raycheva R. 2012). We also do not have enough available data for burnout syndrome in some medical specialties such as specialists working in accident and emergency units (Potter C. 2006). In the specialized literature another analogical research which studies the presence, power and direction of the relations, on the one hand, between the risk factors of the working environment and the psycho-social climate, temperament-personality traits, job satisfaction and on the other, burn out syndrome occurrence, with its psychological dimensions in risk group healthcare specialists, was not found.

The grounds for writing the present chapter of the book are prompted by the necessity for analysis of our own empirical data and their discussion with results from foreign research of the studied problem which would shed more light on the problem (Awang MI. 2012, Koys, DJ. 2001, Maslach C. 1998).

In the result analysis we based our arguments on the fundamental paradigms and contemporary constructs for the subjective well-being and job satisfaction, character-temperament profiles and different models of professional stress and burnout syndrome (Bonev I . 2010, Petkova M. 2003, Cloninger CR. 2011, Koys, DJ. 2001, Maslach C. 1998, Maslach C. 1982).

5.2 Materials and Methods

The object of study of this chapter is a heterogeneous group which consists of several subgroups: **physicians** with specialty in emergency medicine, surgery, anesthesiology and reanimation; nurses with different profile of specialization and educational qualification. A unifying factor of the studied target group is their workplace: emergency medical care aid, surgery and intensive care units.

The distribution of the exerpt according to gender reveals that the women in the present study are 74.3% (n = 75), and the men are 25.7% (n = 26) from a total number of people studied n=101. The age in the monitored group varies within the limits from 23 years to 65 years (average age 42.35 ± 1.07).

The grounds for analyzing the results of the group as a whole and not within the group is the small number of people studied in some of the subgroups and the similar manner of grouping of respondents in analogical foreign studies. Therefore the conclusions refer to the whole target group.

The design of the study, the methods used and validation of the tools are thoroughly described in a separate chapter of the book.

5.3 Results and Discussion

With reference to the age structure, the excerpt has regular distribution of age and a statistically significant difference in age between men and women ($P > 0.05$) was not established. The length of service of the studied group varies from 1 to 41 years (average length 18.18 ± 1.07 years), without statistically significant difference between the two sexes (P > 0.05).

5.4 Personality in the Context of Professional Burnout

It is well-known that temperament traits are inherited and automatically respond to emotional stimuli, which are evident as early as childhood, and remain independent from one another and unaltered throughout one's life. In contrast to them, the character dimensions tend to be modified, i.e. to be influenced by personal experience (Dzamonja-Ignjatovic T. 2010, Maffasioli DG. 2010).

Approximately 10 years later, in one of his studies the researchers S. Mitra, P. K. Sinha et al. reach the conclusion that surgeons and anaesthesiologists have similar personality profiles, i.e. statistically significant differences between the 4 temperament and 3 character dimensions (TCI-125) (personality profile) of 47 surgeons and 46 anaesthesiologists (Mitra S. 2003).

For our study, in the analysis of character and temperament peculiarities of the personality of the studied group, we used the Cloninger model (TCI-R), which is validated for Bulgaria from the authors of the present book (Tilov B. 2010).

The summarized results for the target group studied with TCI-R are presented in Table 5.1.

When comparing the average values of the separate gender scales, a statistically significant difference was not determined with exception of the subscale 'self-directedness' in which men have higher average values (127.58 ± 2.71 v/s 121.12 ± 1.77; P = 0.052) .

Table 5.1 Average and Standard Deviations- Temperament- Character Questionnaire (Revised Version)

TCI-R	N	Number of questions	mean ± SE	SD
Novelty seeking (NS)	101	35	107.74 ± 1.35	13.55
Harm avoidance (HA)	101	33	99.67 ± 1.36	13.66
Reward dependence (RD)	101	30	100.36 ± 1.02	10.34
Persistence (P)	101	35	124.40 ± 1.43	14.37
Self directedness (SD)	101	40	122.78 ± 1.51	15,17
Cooperativeness (C)	101	36	119.20 ± 1.29	12.90
Self-transcendence (ST)	101	26	82.13 ± 1.22	12.26

In another study, the temperament and character personality profile of anaesthesiologists reveals that they have lower average values of 'cooperativeness' and 'persistence', in comparison with physicians from other specialties (Klugel MT. 1999).

The research directed specifically at the influence of the personality profile in burnout occurrence in these risk groups of medical specialists is, unfortunately, insufficient (Bohle A. 2001, Cloninger CR. 2011, Dzamonja-Ignjatovic T. 2010). Authors from a neighbouring country discover the dependence between the levels of burnout syndrome (BS) in surgeons and their temperament- character personality dimensions. Positive correlation is found between the low 'personal achievements' (PA) from MBI and 'reward dependence', 'persistence', 'self-directedness' and 'cooperativeness' and negative correlation between PA and 'harm avoidance' (TCI-240) (Cloninger, CR. 2011).

Since norms for the values of the indicators of TCI-R do not exists, only by empirical data from different populations can we determine where the average values of the participants in the inquiry stand, compared to others. Comparing the average values of the personality profile from the target group with similar other studies, we found that the index 'harm avoidance' is higher in the group studied by us (99.67 $\pm$ 1.36), accordingly: in Serbia- 88.11 for the general population and in the Check Republic- 91.7 (Cloninger CR. 2011, Pejušković B. 2011).

The high values of the index 'harm avoidance' are defined rather like a trend of behavior for repulsing different stimuli or threats. The results give us the grounds to reach the conclusion that the people participating in the inquiry are rather pessimistic, cowardly, indecisive, shy, introvert and emotionally 'worn out' which inevitably influences the distribution and levels of the burnout syndrome experienced by them. In addition to this statement is the fact that the presence of more determined, audacious and stable personal predisposition reduces stress and protects from burnout syndrome (Lyochkova M. 2004).

The opposite dependence is evident, namely that the presence of positive expectations reduces the probability for developing different negative psychic manifestations including burnout syndrome and increases the personal susceptibility to stressogenic events (Petkova M. 2003). In one of her personal studies of subjective well-being, M. Petkova reveals that physicians have higher level of negative expectations and lower level of optimism in comparison with the students, respectively average values of negative expectations - 26.12 and 20.73 and of optimism – 33.58 and 38.17 ((Petkova M. 2003).

The risk from 'burnout' is twice as high with physicians who aim at professional career and experience the lack of free time and support for their development (Embriaco N. 2007).

Results from the studied population (Latin American) in Brazil, working in the system of education and healthcare, reveal the following average values: (NS) 98.8, (HA) 96.8, (RD) 102.5, (PS) 120.6, (SD) 143.9, (CO) 135.8, (ST) 78.3 (Maffasioli DG. 2010). Compared to the seven character-temperament dimensions of the target group studied by us, they indicate that all those characteristics have higher average values with exception to 'self-directedness' and 'reward dependence', i.e. the two protective characteristics from TCI-R for burnout syndrome while 'self-directedness' and 'reward dependence' are not sufficiently expressed.

According to R. Cloninger the dimension 'self-directedness' plays a crucial role on the self-evaluation of health, including satisfaction from life (Brand S. 2010).

The analysis of the results from our study indicates a relatively low self-evaluation of health and job satisfaction, and unsatisfactory self-evaluation of subjective well-being among the medical specialists, participating in the inquiry. The high values of 'cooperativeness' and 'self- transcendence' of the inquired medical specialists in comparison to the same indexes from other studies, reveal a tendency for social tolerance, sympathy, social support and aptitude to experience joy from life. They, on their part, have indirect influence on the subjective well-being and happiness of medical specialists.

This reasserts the profile of the medical specialists, working in accident and emergency units, intensive care units and surgical clinics, which increases the risk from burnout occurrence. In such cases of high level of individual threshold for stressogenic factors, the risk from burnout development if high.

5.5 Assessing Burnout in the Target Group

The summarized results for the burnout level of the studied group of healthcare specialists, assessed with the Maslach questionnaire are presented in Table 5.2.

Statistically significant differences between men and women for the scales 'depersonalization 'and 'personal accomplishment' were not determined. For the subscale 'emotional exhaustion' we found higher average values in women (24.41 $\pm$ 1.23 v/s 19.69 $\pm$ 2.45, P=0.05). Other studies also present results indicating that women suffer more frequently from signs of professional 'burnout' in comparison with their colleagues men (Cloninger CR. 2011, Embriaco N. 2007, Radostina K. 2010).

Table 5.3 presents the results from the frequency distribution of the studied group according to the severity/degree of burnout syndrome for the separate subscales EE, DP and PA.

With respect to the 'emotional exhaustion' scale, the highest relative share is of healthcare specialists with high level of EE (36.76%), followed by those with average level (33,82%) and low (29,42%) level. The distribution of the results in the 'depersonalization' scale is quite the opposite- 7,35% of the participants in the inquiry are with high level, 25.0% are with average level and 67, 65% are with low levels.

In the 'personal accomplishment' scales, the results of the studied group professionals are respectively: 29, 41% with high values, 39, 71% of the respondents are with average levels of physical ability and 30, 88% with values close to the low ones.

This fact is important and should be taken into account when assessing and evaluating the psychic potential of the medical specialists monitored by us. This is evidence for, on the one hand, low subjective job satisfaction and unsatisfactory results from the personal labour performance among the respondents, and on the other, preserved social tolerance, empathy and care for the patient, motivation and search for novelties in the medical science and practice.

When comparing the average results of the subscales (MBI) of the target group, an interesting fact is that the medical specialists have preserved their motivation for work despite the average levels of **'emotional**

Table 5.2 Average and Standard Deviations- Subscales of the Research Instrument of C. Maslach for Diagnosis of Burnout Syndrome (n=101)

MBI	Number of questions	Min	Max	mean ± SE	SD
Emotional exhaustion (EE)	9	3	54	23.20 ± 1.12	11.30
Depersonalization (DP)	4	0	18	6.76 ± 0.44	4.50
Personal Accomplishment (PA)	8	18	46	33.16 ± 0.56	5.63

Table 5.3 BOS Components by Level

BOS	EE		DP		PA	
level	n	%	n	%	n	%
low	30	29, 4	68	67, 6	31	30,9
moderate	34	33,8	25	25,0	40	39,7
high	37	36,8	8	7,4	30	29,4

exhaustion'. This is evidence for the mental stability and personality preservation of the employees in the accident and emergency units, intensive care units and surgical departments. The greatest relative share of the respondents with low levels of **'depersonalization'** presumes the presence of motivation for professional development and humane attitude towards the patients.

The study of Tamara Dzamonja reveals that the result (BS) of the total score of the subscales and average values of DP are not influenced by gender of the studied surgeons. However, statistically significant differences are found in regards to gender in the other two subscales of MBI- EE and PA. Emotional exhaustion is higher in women, while low physical ability (or lack of personal labour achievements) predominates among men (Cloninger CR. 2011).

The level of burnout syndrome (BS) in studied surgeons in Serbia indicates values close to ours (EE – 22.57 (0–27); DP – 5.0 (0–13); PA – 37.27 (0–39) (Cloninger CR. 2011).

According to a study of C. Màslach and S. A. Jackson, physicians who experienced burnout syndrome show higher levels of EE and PA than the control group despite the relatively low levels of depersonalization (Maslach Chr. 1982).

A study among Canadian physicians, working in emergency care, ascertains that the average levels of the assessed subscale of MBI are: EE - 46%, DP - 93% and PA - 79%. Although medical specialists in our study experience considerably greater emotional exhaustion (66% indicate average to high level), they have preserved their humane attitude and empathy for patients. The self-evaluation of their physical ability or personal labour achievements is high in comparison with the group from Canada (Lloyd S. 1994).

As a conclusion, we could summarize that the high levels of 'harm avoidance' (HA) and 'self-transcendence' (ST) from the temperament- character traits, play an significant role in the high EE levels and the risk for burnout syndrome incidence. The results give us the grounds to conclude that the inquired are rather pessimistically disposed, introvert and emotionally 'worn out', which inevitably exerts influence on the occurrence and levels of the burnout syndrome experienced by them. To continue this statement is to state that the presence of more determined, audacious and stable personal predisposition reduces stress and protects from burnout syndrome.

On the other hand, positive expectations reduce the probability from development of different negative psychic manifestations, including burnout syndrome and increases the personality stability towards stressogenic events.

References

[1] Bonev I, V. Boneva. Interrelations between burnout syndrome and job satisfaction in staff in onclogical healthcare clinic. Healthcare policy and management, 2010;4:54–62.

[2] Lyochkova M, M. Todorova, G. Petrova. Attendance of oncological patients as a stressogenic agent in the professional activities of the specialists in health care. Psychosomatic medicine, 2004;1–2:54–59.

[3] Nenova A, A. Rasheva, B. Tsenova, K. Daskalov, I.Marinov. Emotional exhaustion(burnout) and general featurs of the work of medical staff in conservative treatment clinics. Coll. III National Congress of Psychology, Sofia, 28–23 Oct., Sofi-R, ISBN 954-638-137-3, 2005; 316–322.

[4] Petkova M. Subjective well-being and health. PH Kota, S. Zagora, 2003.

[5] Pertova G, M. Lyochkova, N. Mateva, M.Todorova. Psychosocial dimensions of work-related stress in nurses working with oncological patients. Health Care, 2003;2:45–47.

[6] Todorova M, G. Petrova, N. Mateva. Influence of the professional environment in work with oncological patients. Psychosomatic medicine, 2005;2:92–96.

[7] Shopov D, B. Tornyova, A. Raykova. Psychoclimate in hospital departments as a factor for quality enhancement of medical care. General Medicine, 2009;11(1):30–34.

[8] Awang MI, MF Dollard, J.Coward, C. Dormann. Psychosocial safety climate: Conceptual distinctiveness and effect on job demands and worker psychological health. Safety Science, 2012; 50: 19–28.

[9] Balch CM, T Shanafelt. Combating stress and burnout in surgical practice: a review, 2010; 44:29–47.

[10] Bohle A, M Baumgrtel, ML Gotz, EH Muler, D.Jocham. Burn-out of Urologists in the Country of Schleswigholstein, Germany: A Comparison of Hospital and Privite Practice Urologists. The Journal of Urology, 2001; 16(4): 1158–1161.

[11] Brand S, Holsboer-Trashler E. The burnout syndrome-an overview. Ther Umsch, 2010:67(11):561–565.

[12] Cloninger, CR, AH Zohar. Personality and the perception of health and happiness, Journal of Affective Disorders, 2011;128 (1–2):24–32.

[13] Dzamonja-Ignjatovic T, Svrakic D, Svrakic N, Jovanovic M, Cloninger R. Cross-cultural validitation of the evised Temperament and Character Inventory: Serbian data. Comrehensive Psychiatry, 2010;51: 649–655.

[14] Embriaco N, L Papazian, N Kentish-Barnes, F Pochard, E Azoulay, Burnout syndrome among critical care healthcare workers Curr Opin Crit Care, 2007;13: 482–488.
[15] Estryn-Behar M, et al. Emergency physicians accumulate more stress factors than other physicians results from the French SESMAT study. Emerg Med J, 2010; DOI:10.1136/emj.2009.082594.
[16] Garcia-Izquierdo M, Rios-Risquez MI. The relationship between psychosocial job stress and burnout in emergency departments: An exploratory study. Nurs Outlook, 2012:322–9.
[17] Gundersen L. Physician Burnout. Annals of Internal Medicine, 2001; 135(2):145–148.
[18] Hansen N , M Sverke, K Naswall, Predicting nurse burnout from demands and resources in three acute care hospitals under different forms of ownership: A cross-sectional questionnaire survey, International Journal of Nursing Studies, 2009; 46: 96–107.
[19] Känel R. The burnout syndrome. Praxis, 2008; 97(9): 477–87.
[20] Keller KL, Koenig WJ. Management of Stress and Prevention of Burnout in Emergency Physicians. Annals of Emergency Medicine, 1989; 18(1):42–47.
[21] Klimo Jr. P, DeCuypere M, Ragel BT, McCartney S, Couldwell WT, Boop FA. Career Satisfaction and Burnout Among U.S. Neurosurgeons: A Feasibility and Pilot Study. World Neurosurgery 2012; doi: 10.1016/j.wneu.2012.09.009.
[22] Klugel MT, Laidlaw TM, Kruger N, Harrison MJ. Personality traits of anesthetists and physicians: an evaluation using the Cloninger Temperament and character inventory (TCI-125). Anaesthesia, 1999; 54:926–935.
[23] Koys, D.J., DeCotiis, T.A. Inductive measures of psychological climate. Human Relations, 2001; 44: 265–285.
[24] Ksiazek I, TJ Stefaniak, M Stadnyk, J. Ksiazek. Burnout syndrome in surgical oncology and general surgery nurses: A cross-sectional study European Journal of Oncology Nursing, 2011; 15: 347–350.
[25] Lloyd S, D Streiner, S Shannon, Burnout, depression, life and job satisfaction among Canadian emergency physicians, The Journal of Emergency Medicine, 1994; 12(4): 559–565.
[26] Maffasioli D. G., R. Cloninger. Validation and normative studies of the Brazilian Portuguese and American versions of the Temperament and Character Inventory — Revised (TCI-R) Journal of Affective Disorders, 2010; 124: 126–133.

[27] Maslach C, Goldberg J. Prevention of burnout: New perspectives. Applied and Preventive Psychology, 1998; 7(1):63–74.

[28] Maslach Chr. SE Jackson. Burnout in Health Professionals: A Social Psychological Analysis in Social Psychology of Health and Illness, edited by Glenn S Sandera, Jerry Suls, Lawrence Elbraum Associated, ISBN 0-8058-0554-0, 1982.

[29] Mitra S, Sinha PK, Gombar KK, Basu D. Comparison of temperament and character profiles of anesthesiologists and surgeons: a preliminary study. Indian journal of Medical Sciences, 2003; 57(10):431–436.

[30] Pejušković B, Leèić-Toševski D, Priebe S, Toškoviæ O. Burnout syndrome among physicians – personality dimensions and coping strategies. Psychiatria Danubina, 2011; 23(4): 389–395.

[31] Petrova G, Todorova M, Mateva N. Prerequisites for the Occurrence of Burnout Syndrome in Oncology Nurses. Folia Medica, 2005;2: 39–44.

[32] Potter C. To what extent do nurses and physicians working within the emergency department experience burnout: a review of the literature. Australasian Emergency Nursing Journal, 2006; 9:57–64.

[33] Radostina K. Purvanova, John P. Muros. Gender differences in burnout: A meta-analysis. Journal of Vocational Behavior, 2010; 77: 168–185.

[34] Raycheva R, Asenova R, Kazakov D, Yordanov S, Tarnovska T, Stoyanov D. The vulnerability to burn in healthcare personnel according to the Stoyanov-Cloninger model: evidence from a pilot study. International Journal of Person Centred Medicine, 2012; 2(3): 552–563.

[35] Schwartz RW, Barclay JR, Harrell PL, Murphy AE, Jarecky RK, Donnelly MB. Defining the surgical personality: a preliminary study. Surgery, 1994; 115:62–68.

[36] Tilov B, Dimitrova D, Stoykova M, Tornjova B, Foreva G, Stoyanov D. Cross-cultural validation of the Bulgarian language. Journal of Evaluation in Clinical Practice, 2012; 1–6.

6

Burn Out Syndrome Among General Practitioners

Radost Asenova MD, PhD, Associate Professor of General Practice
Department of Health Management,
Health Economics and General Practice,
Faculty of Public Health, MUP

Gergana Foreva MD, PhD, Assistant Professor of General Practice
Department of Health Management,
Health Economics and General Practice,
Faculty of Public Health, MUP

Donka Dimitrova PhD, Engineer, Associate Professor of Health Care Management, Statistical Advisor,
Department of Health management,
health economics and general practice,
Faculty of public health, MUP

Drozdstoj Stoyanov MD, PhD, Full Professor of Psychiatry, Medical Psychology and Person Centered Medicine,
Faculty of Medicine, MUP, Vice Chair Executive Committee PSIG,
Royal College of Psychiatrists, Visiting Fellow, University of Pittsburgh

The current chapter presents the results from the study of the group of general practitioners. The work of the general practitioners is characterized by the use of the holistic approach. GPs take comprehensive care of the patients, which is prolonged, based on the developed trust and collaboration between the doctor and patient, while respecting the patient's autonomy. On the other hand, general practitioners coordinate with medical and non-medical specialists in solving the patients' medical problems. GPs have more contact with patients than any other group of doctors. This large volume of activities, as well as their work schedule and the need for providing continuous care, makes them a vulnerable group for the development of the burnout syndrome.

Drozdstoy St. Stoyanov (Ed.), New Model of Burn Out Syndrome: Towards Early Diagnosis and Prevention, 71–80.

Overall 66 GPs participated in our study. The average age is 46.52±1.06 years and the average work experience – 21.23±1.13 years. The distribution according to sex shows that 72.73% of the people included in the survey are women, which does not allow the search for dependences of the studied categories with respect to men and women.

The aggregate of the study of the individual vulnerability, the measuring of the psychological climate and the evaluation of burnout reveal interrelations specific to the medical specialists working in primary healthcare.

6.1 Temperament and Character Traits of the Personality of GPs According to R. Cloninger's Model (TCR-I)

Cloninger's psychobiological model of personality incorporates 7-factor structure, integrating neurobiological and psychological aspects. This model is based on the idea that personality is determined by temperament and character.

Temperament traits are inherited, while character categories are modelled by personal experience. (Awang 2012)

The data from the TCI-R instruments used with the studied group of GPs are presented in Table 6.1.

High levels of all determinants of temperament could express vulnerability to anxiety, anger and resentment, social rejection and perfectionism. The results obtained for the studied group outline a profile with regard to temperament: ambitious, feeling embarrassed, shy, hesitant, but also enthusiastic, dedicated and warm.

According to Cloninger's psychobiological model of personality, individuals have an outlook of unity when they are highly developed in all three dimensions of character: self-directedness, cooperativeness and

Table 6.1 TCI-R Scores for Individual Dimensions for the Whole Sample, N = 66

Scales	Number of items	Mean ($X \pm$ SE)	(SD)
Temperament			
NS	35	104.41 ± 1.58	12.82
HA	33	96.06 ± 1.51	12.26
RD	30	99.21 ± 1.25	10.16
PS	35	122.39 ± 2.11	17.12
Character			
SD	40	131.20 ± 2.41	19.61
C	36	119.17 ± 1.60	12.95
ST	26	81.05 ± 1.66	13.51

self-transcendence. In terms of character traits the group shows purposefulness, responsibility, creativity, empathy.

This situation is directly related to the high levels of the categories drawn from the study.

These characteristics correspond to the position of the family doctors in the healthcare system, where they are subjected to pressure both from their patients and from other institutions.

The data from studies using the TCI-R instruments in various countries show cultural traits, as well as differences regarding age, sex and profession. (Eley 2009, Snopek 2012)

A study by D. Eley et al. examines 214 GPs. There should be noted the lower values of the average scores indicating both positive and negative aspect along all categories of the scale. The studied group has high values according to the SD and C scales, but lower levels for the ST and PS categories, which is assessed by the authors as expected about the medical profession. (Eley 2009)

6.2 Assessment of the Organizational and Psychological Climate of GPs Utilizing D. Koys and T. De Cottis's Instruments

Organizational and psychological climate is a concept including not only physical factors but also psychological aspects. The assessment of the psychological climate with the used instruments shows as to what extent the environment stimulates or demotivates the employees. (Dzamonja-Ignjatovic 2010)

The data from the questionnaire used with GPs are shown in Table 6.2.

This instrument measures how the environment factors are interpreted by the studied GPs and what meaning is attributed to them in relation to

Table 6.2 IMPC Scores for Individual Dimensions for the Whole Sample, N = 66

Scales	Number of items	Mean (±SE	(SD)
Autonomy	5	20.58 ± 0.42	3.44
Cohesion	5	18.71 ± 0.51	4.11
Trust	5	19.32 ± 0.62	5.01
Pressure	5	16.61 ± 0.55	4.44
Support	5	19.00 ± 0.65	5.26
Recognition	5	17.58 ± 0.44	3.60
Fairness	5	19.03 ± 0.57	4.64
Innovation	5	18.86 ± 0.62	5.06

the temperament and character traits of personality of GPs and the levels of burnout syndrome.

The data from the study reveal positive correlations, with an activity of various strength, between the separate subscales of the used instruments and are presented in Table 6.3.

These results could be a starting point for future studies aiming at a detailed outline of the personality traits of GPs, related to the specific organizational climate in general practice, which is characterized by tow descriptive dimensions – norms and organizational structure, reward and control mechanisms.

As far as the categories age and work experience are concerned, dependencies were established only in the subscales shown in Table 6.4.

Measuring the burnout syndrome of GPs according to C. Maslach's scale.

GPs were identified to be highly affected by burnout syndrome. The average score of the separate subscales are shown in Table 6.5.

In analyzing the categories, we found out that 42.43% of the studied people have a high degree of EE, 18.18% middle degree, and 39.39% are those with a low degree.

Table 6.3 Correlations between the TCI-R Subscales and the IMPC Subscales

			IMPC					
			Norms of an organization			Organizational structure, reward and control mechanisms		
			Trust	Support	Fairness	Recognition	Pressure	Innovation
TCI-R	Temperament	NS				0,24 *		
		HA	0,37 *			0,34 **	0,45 **	
		RD	0,30 *	0,35 **	0,24 *	0,44 **		0,30 *
		PS	0,31 *	0,46 **	0,53 **			0,44 **
	Character	SD						0,25 *
		ST		0,28 *	0,33 **	0,30 *		0,29 *

* Correlation is significant at the 0.05 level.

** Correlation is significant at the 0.01 level.

Table 6.4 Dependencies between Age, Work Experience and the TCI-R and IMPC Scales

	Age	Work experience
TCI-R		
HA	0,27 *	0,29 *
IMPC		
Pressure	0,26 *	0,25 *

* Correlation is significant at the 0.05 level.

Table 6.5 MBI Scores for Individual Dimensions for the Whole Sample, N = 66

Scales	Number of items	Mean ±SE	(SD)
EE	9	23.77 ± 1.52	12.34
DP	4	7.32 ± 0.60	4.89
PA	8	31.97 ± 0.74	5.97

With the DP category, it was found out that the highest percentage of GPs are with a low degree of 53.03%, followed by those with a middle degree of 33.33 % and the lowest percentage of GPs with the highest degree of depersonalization 13.64%.

With the PA category, the analysis of the group showed that 16.67% of GPs are with low levels, which is an unfavourable position. 40.91% are GPs with a middle degree of work efficiency and the highest percentage of GPs are with high indicators – 42.42%.

The reform in healthcare in Bulgaria started in 2000, with the introduction of GPs as the main figure in the healthcare system. Maslach's scale was used for assessing the burnout syndrome in the very first years after the thorough change in the structure of the Bulgarian healthcare system, when the family doctors had to assume their completely new functions for an extremely short period. The study includes 69 GPs, and the data show that the burnout syndrome includes a considerable part of GPs – 62.3% – with a high degree of EE, 24.6% - with a high degree of DP and 15.9% with a low degree of PA. More than half of the studied GPs are with a middle and a high degree of stress according to the general assessment of the three indicators for measuring the burnout syndrome. (Asenova 2003)

As evident from the comparison of the two studies, there is a considerable change of the values of the EE and DP categories, but they still remain high.

The data from the current study show higher levels of burnout syndrome among GPs in comparison with data from studies of the syndrome among other medical specialists. (Arigoni 2010, Kushnir 2006, Virtanen 2008)

In a study by C. Maslach et al. (2001), GPs from South European countries show significantly lower levels of burnout in the EE category, but higher in the category of reduced PA.

A study by Orton 2012 among 789 GPs in the United Kingdom shows high values of EE at 46%, high values of DP at 42% and low values of PA at 34%.

In a huge multinational survey in 12 countries 1393 GPs are studied. 43 % of GPs are with high values of EE, 35.3% high DP, 32.0% low degree of PA, while 12% of the respondents show high level in all three categories. National specifics have been found out. GPs from Bulgaria, Italy and the

United Kingdom are with the highest percentage of EE. The highest levels of DP are observed among GPs in Greece, Italy and the United Kingdom, and reduced PA related to burnout in Greece and Turkey. The study finds out considerable differences in the percentages of GPs in the separate categories in the various countries (from 15 to 68% for high levels of EE, from 12 to 73% for DP, from 12 to 93% for PA and from 2 to 25% high levels of burnout in the three categories). (Soler 2008)

Studies show a clear tendency regarding cultural traits in manifestation of burnout.

Dependencies have been found regarding multiple factors of the environment.

Thus, a number of studies cover and prove the high level of the syndrome at low job satisfaction. (Deckard 1992, Linzer 2001, Maslach 2001, Thommasen 2001)

The doctor who hold an academic position are characterized by burnout syndrome combining reduced work efficiency and low level of emotional exhaustion. (Maslach 2001)

Along with the external factors, personality factors are undoubtedly related to the syndrome. M. Linzer et al. show the relation between manifestations of burnout and the doctors' personal perception of low control over their work environment. (Linzer 2001)

The low self-evaluation has been proven to be related to the presence of burnout. (Maslach 2001)

The important link between burnout and personality was demonstrated in numerous publications by using personality inventories.

In searching for correlations between the three mainly studied components of the theoretical model – personality, environment and burnout of GPs considerable dependencies were found.

Our results using a one-way analysis of variance for a quantitative dependent variable by a single factor (independent) variable show significance between EE and temperament trait PS, character trait SD and P related to organizational climate. The data from the one-way analysis are presented in Table 6.6.

Spearman's correlations were performed to provide the associations between the TCI-R determinants, IMPC dimensions and MBI scores and to measure the strength and direction of each of these associations. The results are presented in the following Table 6.7.

According to the result, a high score of the dimension HA, which expresses vulnerability to anxiety, leads to a high level of EE.

Table 6.6 One-way Analysis of Variance for the Influence of Factors Related with the Personality and Psychosocial Climate, and the Level of Occupational Exhaustion

	MBI			
	PA		EE	
	F	P	F	P
TCI-R				
PS	1.90	< 0.05		
SD	2.00	< 0.05		
IMPC				
Pressure			2.94	< 0.01

Table 6.7 Significant Correlations between MBI and TCI-R, IMPC

	MBI		
	EE	DP	PA
TCI-R			
HA	0,36 **	0,27 *	-0,25 *
PS			0,28 *
IMPC			
Cohesion	0,33 **		
Pressure	0,62 **	0,30 *	

* Correlation is significant at the 0.05 level.

** Correlation is significant at the 0 01 level.

The fact that GPs most often work alone, including in remote areas, isolated in terms of professional contacts with colleagues, could explain the interrelation between C and high levels of EE.

A strong positive correlational dependence is found between the scale EE of MBI and the scales P and C of IMPC, where P< 0.01. The high levels of the EE scale assume that the personality experiences the pressure from the patients, the administrative departments, and the general social and economic situation. The giver dependence provides an opportunity for the use of appropriate strategies, which, by reducing the pressure, will also reduce the manifestations of burnout.

It is noticeable that the DP category of MBI, i.e. a high degree of loss of humanity, desire for work and development, even though manifested at low percentage of GPs, is in a weak positive correlational dependence both on the scale for HA of TCI-R and on the scale for P of IMPC. These results also reveal the potential for prevention of the burnout syndrome, via regulation of the perception of time demands with respect to task completion and performance standards and HA.

It is evident from the results that there is a weak negative dependence between the PA scale of MBI and the HA scale of the TCR-R questionnaire. The low levels of the HA scale presume GPs to be more optimistic, self-confident and more active. There is no doubt that, with personality and temperament traits like anxiety, shyness and hesitancy, work efficiency will be influenced negatively. The high levels of the temperamental characteristic PS, most often express by perfectionism on the part of GPs increase professional development and improvement.

Burnout is a problem of current importance to be studied among family doctors. We found that the most important dimension from personality characteristics is harm avoidance that makes doctors more prone to burnout syndrome. High level is recognized as a factor which leads to the development of burnout. Persistence is assumed to be a protective dimension.

The contribution of our findings is that psychological climate might be very important in the development of burnout syndrome.

The used battery, which consolidates evaluation of individual vulnerability, psychological climate and burnout, reveals correlations that can be employed in the prevention of burnout syndrome by the balance of such factors as harm avoidance by regulating pressure.

Unfortunately, burnout is difficult to prevent, so we must focus on early detection of burnout in general practitioners, rather than later symptoms.

References

[1] Asenova R, Yaman H, Soler JK et all. Burnout among Bulgarian GPs. Ith National Conference of studying stress. Plovdiv 2003, Abstract book:207–213.

[2] Arigoni F, Bovier P, Sappino A. Trend inburnout among Swiss doctors. Swiss Med Wkly. 2010, doi:10.4414/smw.2010.13070

[3] Awang MI, MF Dollard, J.Coward, C. Dormann. Psychosocial safety climate: Conceptual distinctiveness and effect on job demands and worker psychological health. Safety Science 2012; 50: 19–28.

[4] Deckard GJ, Hicks LL, Hamory BH. The occurrence and distribution of burnout amongst infectious diseases physicians. The Journal of Infectious Diseases 1992; 165: 224–8.

[5] Dzamonja-Ignjatovic T, Svrakic D, Svrakic N, Jovanovic M, Cloninger R. Cross-cultural validitation of the evised Temperament and Character Inventory: Serbian data. Comrehensive Psychiatry. 2010;51:649–655.

[6] Eley D, Young L, Przybeck T. Exploring the treatment and character trials of rural and urban doctors. The journal of rural Health. 2009, vol25,No 1:43–49

[7] Kushnir T, Cohen AH. Job structure and burnout among primary care pediatricians. Work. 2006;27:67–74.

[8] Linzer M, Visser MRM, Oort FJ, Smets EMA, McMurray JE, de Haes HCJM, for the Society of General Internal Medicine Career Satisfaction Study Group. Predicting and preventing physician burnout: results from the United States and the Netherlands. The American Journal of Medicine 2001; 111: 170–175.

[9] Maslach C, Schaufeli WB, Leiter MP. Job burnout. Annual Review of Psychology 2001; 52: 397–422.

[10] Orton P, Orton C, Gray D. Depersonalised doctors: a cross-sectional study of 564 doctors, 760 consultations and 1876 petient reports in UK general practice. BMJ Open 2012;2:e000274. doi:10.1136/bmjopen-2011–000274

[11] Snopek M, Hublova V, Porubanova M, Blatny M. Psychometric properties of the Temperament and Character Inventory-Revised (TCI-R) in Czech adolescent sample. Comprehensive Psychiatry 53 (2012) 71–80

[12] Soler JK, Yaman H, Esteva M, Dobbs F, Asenova RS, Katic M, Ozvacic Z, Desgranges JP, Moreau A, Lionis C, Kotányi P, Carelli F, Nowak PR, de Aguiar Sá Azeredo Z, Marklund E, Churchill D, Ungan M; European General Practice Research Network Burnout Study Group. Burnout in European family doctors: the EGPRN study. Fam Pract. 2008 Aug;25 (4):245–65.

[13] Thommasen HV, Lavanchy M, Connelly I, Berkowitz J, Grzybowski S. Mental health, job satisfaction and intention to relocate. Canadian Family Physician 2001, B47: 737–44.

[14] Virtanen P, Oksanen T, Kivimaki M, Virtanen M, Pentti J, Vahtera J. Work stress and health in primary health care physicians and hospital physicians. Occup Environ Med. 2008;65:364–6.

7

Comparative Analysis of Vulnerability to Burnout Syndrome in Health and Social Care Personnel

Ralitsa Raycheva Assistant Professor,
Department of Social Medicine and Public Health,
Faculty of Public Health, MUP

Christina Georgieva MA in Clinical Psychology,
PhD Candidate, Coordinator at Agency for Social Support

The frequency of emotional stress is growing and covers all social groups as a consequence of our modern, dynamic life in a world of information workload, urbanization and conflicts of different nature. Stress is a physiological and psychological process by which an individual responds to events or situations that bring increased demands and exert pressure on him. Any change to which one must adapt, is accompanied by stress. Stress is a process by which we evaluate the advancing environmental changes and respond to the increased behavioral and functional demands on organism or threats to its integrity. Occupational stress occurs with the interaction between people on the field of their professional activities. Negative sources of occupational stress may be associated with both organizational characteristics and processes, working conditions and interpersonal interactions and personality predictors of employee.

This study investigates the associations between personality traits, psychological climate and burnout in health and social care workers.

Two groups, both with 25 randomly selected health and social workers, were drawn from a region of Southern Bulgaria.

The main objectives of the study could be summarized in the following areas – (I) to explore personal predisposition of both groups under research; (II) to trace the psychological climate and its significance for burnout; (III) to identify the forms of manifestation of burnout for the studied groups.

Drozdstoy St. Stoyanov (Ed.), New Model of Burn Out Syndrome: Towards Early Diagnosis and Prevention, 81–106.

Hypotheses of the study – (I) no statistically significant difference is expected between personal disposition of the two groups in the sample with respect to the manifestation of burnout; (II) psychosocial climate in both treatment groups is supposed to be characterized by similar determinants due to the similar working conditions; (II) elevated levels of MBI scales are expected for both groups; (III) high levels of "Pressure" are expected to positively correlate with "Emotional Exhaustion" and negatively with the "Personal Accomplishment".

Tasks of the study - (I) to examine and analyze the accessible literature on the problem of the study; (II) to investigate the association between personality, work environment and burnout in health professionals and social workers; (III) to establish the validity of psychobiological theory and its significance for the phenomena of burnout.

7.1 Methods

7.1.1 Sample

Burnout occurs most often in occupations that require intensive contact with people; this includes professionals in the field of health and social care. Cases in the sample were randomly selected from a region in Southern Bulgaria–Plovdiv region, municipality of Plovdiv and Pazardzhik. The units of observation (n = 50) who participated in the survey work in the area of health and social services – health care professionals (n = 25) and social welfare workforce (n = 25).

7.1.2 Measures / Instruments

7.1.2.1 Temperament and Character Inventory

TCI is a self-report measure of personality based on the theory proposed by Cloninger et al. [1]. We decided to employ the 5-point Likert scale (1. definitely false; 2. mostly false; 3. not true, nor false; 4. mostly true & 5. definitely true), 240 items TCI – revised version (TCI-R). The response format of TCI-R increases the reliability of the responses because graded responses are possible.

7.1.2.2 Inductive Measurement of Psychological Climate (IMPC)

The test of Koys and DeCotiis is a self-reported measure with 8 categories of psychological climate perception: Cohesion, Trust, Support, Fairness, Autonomy, Pressure, Recognition and Innovation [2]. In this study, we used 45 of the original 80 criteria that were retained for describing the climate.

Responses are rated according to a 5-class Likert scale. Participants were asked to rate how each description corresponded to their own perception of the working environment.

7.1.2.3 Maslach Burnout Inventory (MBI)

MBI is a self-reported measure of burnout syndrome that weighs the effects of emotional exhaustion and reduced sense of personal accomplishment [3]. The MBI consists of 22 items yielding scores for three components – Emotional exhaustion (EE), Depersonalization (D) and Personal Accomplishment (PA). The participants were asked to provide a rating with a 7-class scale for each item. Its proper application enables the analysis of the consequences of the professional burning process along with analysis of the psycho-emotional environmental effects. However, in isolation it may turn out to be insufficient in terms of the evaluation of numerous details and specific characteristics of every contingent studied.

7.1.3 Procedure

The questionnaire [Battery of Assessment Tools (BAT) – TCI-R, IMPC, and MBI] was administered individually to the subjects. At the beginning of administration, necessary information was presented. The health and social care workers completed the questionnaires during their work hours and returned them directly to the researchers. The cover page of the BAT included questions about demographics. To protect privacy, the questionnaires were anonymous and no personal information was requested.

7.1.4 Statistical analysis

Data analysis was performed using analytical and descriptive statistical methods, and multivariate statistical models. Prior to statistical analyses, all of the variables were examined for: accuracy of data entry, missing values and fit between their distributions and assumptions of multivariate analysis. The hypothesis of the distribution's normality of the analyzed variables was tested using the Kolmogorov-Smirnov test. Demographics were evaluated by descriptive statistical methods. Spearman's rank correlation coefficients were calculated among scores on the TCI-R, IMPC and MBI. Spearman's correlations were performed to provide the associations between the TCI-R determinants, IMPC dimensions and MBI scores and to measure the strength and direction of each of these associations. For all variables we proved normal distribution and performed comparison between study groups with analysis of variance (one-way ANOVA). In all performed analyses, a significance criterion equal to or smaller than 0.01, was used to determine statistical significance, except for two tests where the adopted significance levels were $p < 0.05$ and $p < 0.001$. The results were processed with professional statistical package for social studies SPSS - v.19, programming software MATLAB - v.2011a and professional graphic design software Photoshop - v.5.1.

7.2 Empirical Studies

7.2.1 Demographics

The research sample consisted of 50 employees, equally distributed to health and social care professionals. Their mean age was 44.8 (SD±11.8, range 24–68). Most participants in the sample consisted of females (70%); worked in 13 specialized institutions for medical and social activity form which State-run Psychiatric Hospital Pazardzhik has the biggest share (20%), followed by those employed in the Daycare Facilities for Disabled Children – 16% and a Daycare Facilities for Disabled Adults – 14%; had an average length of service of 19.9 years (SD±12.58, range 1–43) and an average organization tenure was 5 years (SD±4.52, range 3m-20y). Furthermore, the sample was primarily comprised of nurses (16%), followed by 10% for psychiatrists and social care workers and sixteen other employee classifications (64%).

7.2.2 Statistical Processing of the Results Obtained from BAT

The values of the mean and the standard deviation (n = 50) for the TCI-R, IMPC and MBI were respectively 757.62 (±SD 70.57), 144.76 (±SD 23.84) and 60.22 (±SD 20.95). On average, the sample was composed of subjects reporting low/moderate/high scores of Temperament and Character Inventory; low/moderate/high scores of psychological climate; and a low/moderate/high score of burn out syndrome (see Tables 7.1–7.3). In order to categorize out results, TCI-R normative data, Maslash's MBI-Scoring key and 5 (number of the questions in each group of psychological climate's 8 categories) x 5 (5-point scale) multiplications for IMPC were used [4].

7.2.3 Statistical Analysis of the Associations between the Scales in BAT

We applied correlation analysis in order to empirically confirm the hypothesis – the existence of a specific interaction between temperament and character's

Table 7.1 Means and Standard Deviations for Temperament and Character Inventory – Revised Subscales

Scales	Temperament Scales	n = 50		
		Items	Mean	± SD
TCI-R		240	757.62	70.57
NS	Novelty Seeking	35	108.20	11.80
HA	Harm Avoidance	33	98.02	11.72
RD	Reward Dependence	30	99.28	10.66
PS	Persistence	35	119.00	16.96
Scales	Character Scales	n=50		
		Items	Mean	± SD
SD	Self-Directedness	40	123.24	17.59
CO	Cooperativeness	36	115.22	12.03
ST	Self-Transcendence	26	79.58	12.76

Table 7.2 Means and Standard Deviations for Psychological Climate Dimensions

Scales	Inductive measurement of psychological climate	n=50		
		Items	Mean	± SD
IMPC		40	144.76	23.84
Autonomy (A)		5	0.83	4.04
Cohesion (C)		5	16.88	4.42
Trust (T)		5	20.16	4.60
Pressure (P)		5	16.16	3.22
Support (S)		5	19.12	4.92
Recognition (R)		5	16.92	3.52
Fairness (F)		5	18.64	4.23
Innovation (I)		5	18.14	5.03

Table 7.3 Means and Standard Deviations for Maslach Burn-out Inventory Subscales

Scales	Maslach Burnout Inventory	n=50		
		Items	Mean	± SD
MBI		22	60.22	206.95
Emotional exhaustion (EE)		9	22.78	10.08
Depersonalization (D)		4	6.12	4.18
Personal Accomplishment (PA)		8	31.90	5.70

Table 7.4 Correlations between Harm Avoidance (TCI–R), Trust and Pressure(IMPC) and Emotional Exhaustion – EE (MBI)

Spearman's rho (n=50)		Trust	Pressure	EE
Harm Avoidance	Correlation Coefficient	.018	.296*	.133
	Sig. (2-tailed)	.900	.037	.358
Trust	Correlation Coefficient		.182	–.006
	Sig. (2-tailed)		.206	.970
Pressure	Correlation Coefficient			.426**
	Sig. (2-tailed)			.002

*** Correlations with p-values < 0.001.

** Correlations with p-values < 0.01.

* Correlations with p-values < 0.05.

determinants, psychological climate's dimensions and burnout syndrome's domain. The correlation is a mathematical term used in a general sense as a measure of stochastic (probabilistic, non-functional) dependence between random variables. Milestone in the implementation of the correlation analysis is the choice of an appropriate correlation coefficient. In this study, the results of the data processing is represented by the Spearman's rank correlation coefficient (rho-ρ), which is a valuable tool for establishing association and estimation of the power between two, usually paired, groups that initially appear in rank order. The following tables summarize the correlations in pairs and give information about the extent and direction of the association between variables.

Table 7.4, shows a low positive correlation between Harm Avoidance and Emotional Exhaustion (ρ=0.296, P=0.037); and moderate association between Pressure and Emotional Exhaustion (ρ=0.426, P=0.002). In summary, high scores in Harm Avoidance in combination with Pressure are associated with high scores of Emotional Exhaustion that lead to burn out syndrome.

The results presented in the matrix of **Table 7.5** demonstrate low and moderate positive correlation between "Autonomy" and "Support" (ρ = 0.302, P = 0.033) and "Innovation" and "Support" (ρ = 0.552, P = 0.000).

The results for the correlation between one temperamental dimension, two categories of psychological climate and two components of burn out manifestation are summarized in **Table 7.6**. In total, 2 significant correlations were obtained. Persistence had a moderate positive correlation with Autonomy (ρ=0.432, P=0.002) in psychological climate domain and

Table 7.5 Correlations between Novelty Seeking (TCI–R), Autonomy, Innovation, Support (IMPC) and Personal Accomplishments (MBI)

Spearman's rho (n=50)		Autonomy	Innovation	Support	PA
Novelty Seeking	Correlation Coefficient		–.115	–.034	.020
	Sig. (2-tailed)	.112	.427	.817	.891
Autonomy	Correlation Coefficient		.267	.302*	.086
	Sig. (2-tailed)		.060	.033	.554
Innovation	Correlation Coefficient			.552***	–.147
	Sig. (2-tailed)			.000	.309
Support	Correlation Coefficient				.080
	Sig. (2-tailed)				.579

*** Correlations with p-values < 0.001.

** Correlations with p-values < 0.01.

* Correlations with p-values < 0.05.

likewise with Personal Accomplishments (ρ=0.442,P=0.023) in burnout domain.

Table 7.7 shows significant positive correlation between Recognition and Support (ρ=0.613, P=0.000). Generally, low scores in Recognition and Support are associated with high Emotional Exhaustion. In this sample distribution the expected moderate or high negative associations between the scales of the "Recognition" and "Support" in the psychological climate domain with the scale of "Emotional Exhaustion" in burnout domain were not proven.

Table 7.8 illustrates multiple significant associations. Within the dimensions of psychological climate a moderate positive correlation exist between "Cohesion" and "Trust" (ρ = 0.460, P = 0.001) on the one hand, and "Fairness" on the other (ρ = 0.400, P = 0.004). Strong positive association was observed between "Trust" and "Fairness" (ρ = 0.846, P = 0.000). In the domain of burnout syndrome significant correlation occurs between "Depersonalization" and "Emotional Exhaustion" (ρ = 0.575, P = 0.000).

Table 7.9 represents the associations between the variables as follows – moderate, with positive direction between "Self-Directedness" with "Autonomy" (ρ=0.442, P=0.001) and with "Emotional Exhaustion" (ρ=0.338, P=0016).

Table 7.10 summarizes the correlations between the scales in each of the three instruments –TCI-R, IMPC, MBI, and demonstrates the strength and direction of the associations between scales throughout the battery (see the Appendix).

Table 7.6 Correlations between Persistence (TCI–R), Autonomy, Innovation (IMPC), Personal Accomplishments and Emotional Exhaustion (MBI)

Spearman's rho (n=50)		Autonomy	Innovation	PA	EE
Persistence	Correlation Coefficient	.432**	.139	.442**	.129
	Sig. (2-tailed)	.002	.337	.023	.373
Autonomy	Correlation Coefficient		.267	.020	.050
	Sig. (2-tailed)		.060	.892	.732
Innovation	Correlation Coefficient			–.102	.128
	Sig. (2-tailed)			.482	.374
Personal Accomplishment	Correlation Coefficient				–.245
	Sig. (2-tailed)				.087

*** Correlations with p-values < 0.001.

** Correlations with p-values < 0.01.

* Correlations with p-values < 0.05.

Comparative analysis was conducted on each of the indicators gender, age, length of service and length of service at last workplace for each of the subscales of the TCI-R, IMPC, MBI. Statistically significant differences were found in the following cases:

Table 7.7 Correlations between Reward Dependence (TCI–R), Recognition, Support (IMPC) and Emotional Exhaustion (MBI)

Spearman's rho (n=50)		Recognition	Support	EE
Reword Dependence	Correlation Coefficient	-.023	-.092	.021
	Sig. (2-tailed)	.875	.524	.887
Recognition	Correlation Coefficient		.613***	.201
	Sig. (2-tailed)		.000	.162
Support	Correlation Coefficient			.053
	Sig. (2-tailed)			.716

*** Correlations with p-values < 0.001.

** Correlations with p-values < 0.01.

* Correlations with p-values < 0.05.

Table 7.8 Correlations between Cooperativeness (TCI–R), Cohesion, Trust, Fairness (IMPC), Depersonalization and Emotional Exhaustion (MBI)

Spearman's rho (n=50)		Cohesion	Trust	Fairness	D	EE
Cooperativeness	Correlation Coefficient	.043	.131	.137	–.038	.093
	Sig. (2-tailed)	.769	.364	.344	.792	.521
Cohesion	Correlation Coefficient		.460**	.400**	–.052	.053
	Sig. (2-tailed)		.001	.004	.719	.715
Trust	Correlation Coefficient			.846***	–.150	–.006
	Sig. (2-tailed)			.000	.299	.970
Fairness	Correlation Coefficient				–.228	–.001
	Sig. (2-tailed)				.112	.992
Depersonalization	Correlation Coefficient					.575***
	Sig. (2-tailed)					.000

*** Correlations with p-values < 0.001.

** Correlations with p-values < 0.01.

* Correlations with p-values < 0.05.

Table 7.9 Correlations between Self-Directedness (TCI–R), Autonomy, Innovation (IMPC), Personal Accomplishments and Emotional Exhaustion (MBI)

Spearman's rho (n=50)		Autonomy	Innovation	PA	EE
Self-Directedness	Correlation Coefficient	.442**	.172	.036	.338*
	Sig. (2-tailed)	.001	.232	.801	.016
Autonomy	Correlation Coefficient		.267	.078	–.017
	Sig. (2-tailed)		.060	.591	.907
Innovation	Correlation Coefficient			–.147	.128
	Sig. (2-tailed)			.309	.374
Personal Accomplishments	Correlation Coefficient				–.150
	Sig. (2-tailed)				.299

*** Correlations with p-values < 0.001.

** Correlations with p-values < 0.01.

* Correlations with p-values < 0 05.

gender was associated with one scale of IMPC - “Fairness” (F=6.72, P=0.013);
age was associated with four of the eight scales of IMPC - “Trust” (F=2.58, P=0.016), “Support” (F=2.59, P=0.015), “Recognition” (F=2.19, P=0.036) , “Fairness” (F=2.67, P=0.013);
length of service was associated with two of the scales of TCI-R - “Reward Dependence” (F=2.30, P=0.025) and “Harm Avoidance” (F=2.69, P=0.010);
length of service at last workplace was associated with one of the TCI-R’s and one of the IMPC’s scales – “Reward Dependence” (F=2.17, P=0.035), and “Recognition” (F=2.46, P=0.018) .

We proved the association between “Reward Dependence”, “Persistence” (as dimensions of temperament), “Pressure” (as determinant of the psychological climate) and the level of “Personal Accomplishment” (as a characteristic of burnout syndrome) by applying one-way ANOVA analysis. We found significant associations between the dimensions of “Pressure”, “Recognition” and “Innovation” with “Emotional Exhaustion” – for its absence or presence. Person under the influence of “Pressure”, who experienced lack of “Recognition” and the impossibility of “Innovation” in the work environment, is destined to burnout. “Recognition” was an important motivational trait for effectiveness and survival of the personnel in the field of medicine and social care. “Pressure” is an extremely important element of the working environment, which in most cases has a tangible presence in everyday working activities. The positive or negative impact of the psychological climate on personality profile of the individual depends on which kind of temperament and character determinants were combined with the dimensions of the working environment. In our case “Pressure” coupled with “Reward Dependence” and “Persistence” has a beneficial effect in a competitive environment and increase ambitions for better „Personal Accomplishment”. Although in combination with a hostile work environment, without motivation, driven by a lack of “Recognition” and “Innovation”, “pressure” appears to be a major factor for the occurrence of “Emotional Exhaustion”. The analysis did not show association between variables with the third characteristic of burnout syndrome – “Depersonalization” (Table 7.11).

7.2.4 Analysis of the Sample by the type of Activity – Medical and Social

In this part we shall compare two groups of specialists - professionals employed in health care and social welfare.

The first group consists of health professionals: doctors, nurses and therapists. Doctors represent 72% of the units of observation in this group. They

Table 7.10 One-Way ANOVA Analysis: Association between both TCI-R and IMPC Variables with the Level of Burnout Syndrome

	MBI					
TCI-R, IMPC	Emotional Exhaustion		Depersonalization		Personal Accomplishment	
	F	P	F	P	F	P
Reward Dependence					3.13	0.003
Persistence					2.48	0.013
Pressure	3.93	0.001			2.13	0.031
Recognition	3.00	0.005				
Innovation	1.95	0.050				

are a heterogeneous group from different medical fields with the following proportions – 23.53% Psychiatry; 17.64% General Medicine and Surgery; 11.76% Clinical Psychology; and 5.89% Pediatrics and Internal Medicine, Neurology, and Orthopedics and Traumatology. Nurses represent 24% of the sample and Rehabilitators, respectively 4%. Health professionals perform their activities in a total of 8 healthcare facilities.

The second group consists of social welfare workforce with the following proportions – 20% Social Activities; 16% Education; 12% Psychology; 8% Nurse, Preschool and Elementary Education, Psychology, Social Pedagogy; 4 % Engineer, Health Management, Preschool and Social Education, Preschool education, preschool education with social pedagogy. Social welfare workforce performs their activities in 5 institutions.

Women accounted for approximately 68% of the overall sample in the first group and 72% in the second. For the first group respondent age ranged from 27 to 68 with a mean sample age of approximately 46 years (SD±12.57), for the second respectively from 24 to 58 with a mean 43 years (SD±11.255).

Proportions of respondents' length of service and organization tenure are presented in **Table 7.12**, respectively, for medical and social activities. Mean values for length of service and organization tenure are presented in **Table 7.13**.

The distribution by gender in both groups did not allow establishing the significance between men and women. Significant differences in mean scores between health and social care personnel were not proven by age, length of service and organization tenure.

The results of the average score of the TCI-R scales did not show statistically significant differences between the two groups in any of the seven scales. The lack of significance did not allow us to frame comparative

Table 7.11 Proportions of Respondents by Length of Service and Organization Tenure

	Group I – Healthcare professionals		Group II – Social welfare workforce	
Years	Length of service	Organization tenure	Length of service	Organization tenure
up to 5 years	24%	56%	12%	64%
from 5 to 10	8%	32%	28%	24%
from 11 to 20	12%	8%	4%	12%
from 21 to 30	36%		32%	
over 30 years	20%		24%	

Table 7.12 Mean Values for Length of Service and Organization Tenure

Group I – Healthcare professionals			
	N	Mean	±SD
Length of service	25	20.28	12.525
Organization tenure	25	4.68	3.518
Group II – Social welfare workforce			
Length of service	25	19.52	12.884
Organization tenure	25	5.36	5.390

personality profile of the two groups of specialists. Hence the assumption that the temperament and character profile of respondents matches, due to the homogeneity of the sample, was is valid.

The results of the IMPC analysis between the two groups showed no statistically significant difference in the field of psychological climate of medical and social employees. Therefore the suggestion that the determinants of the work environment in the medical and social field are similar and have the same impact on employment was completely reasonable.

The surveyed specialists from both groups have high scores in terms of "Trust", "Support", "Autonomy", "Fairness" and "Innovation". Social welfare workforce gave higher scores on all the psychological climate characteristics, except for "Pressure", but the difference was not significant (Figure 7.1).

There was not statistically significant difference between the personnel, employed in medical and social field, regarding the three scales of burnout syndrome – "Emotional Exhaustion", "Depersonalization" and "Personal Accomplishment". The units of observation from the two researched groups were distributed depending on the score of the scales– high, moderate and low. We used the specified MBI Scoring Key (results key):

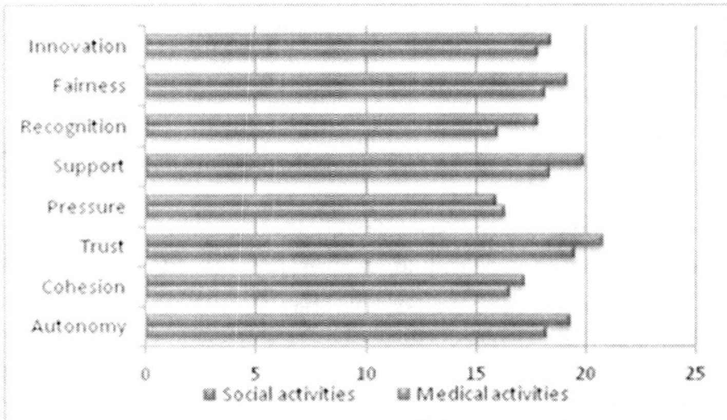

Figure 7.1 Proportions of IMPC Scales in both Researched Groups – Health and Social Care Activities.

"Emotional Exhaustion"

- o Low - from 0 to 16 points
- o Moderate - 17 to 26 points
- o High - over 27 points

"Depersonalization"

- o Low - 0 to 6 points
- o Moderate - 7 to 12 points
- o High - over 13 points

"Personal Accomplishment" (the ranking scale is reversed)

- o Low - over 39 points
- o Intermediate - 32 to 38 points
- o High level - from 0 to 31 points

The mean values of the scales show that respondents in both groups had a moderate manifestation of burnout syndrome. The mean (±SD) of the scale "Emotional Exhaustion" for social welfare workforce was 21.56 (±11.04) and 19.64 (±10.82), respectively, for the health care professionals (i.e. both groups have moderate score – 17–26 points). Social welfare workforce performs moderate level of "Depersonalization" – 7.12 (±4.00), while health care professionals fall within the low level of expression – 5.12 (±4.19). Both groups show moderate results for "Personal Accomplishment" – 29.80 (±4.43) for social and 29.64 (±5.47) for health care activities.

We perform analysis of the frequency distribution of the researched groups according to the three subscales levels (Figures 7.2 and 7.3).

Alarming was the result that over 50% of respondents in both groups had high levels of "Emotional Exhaustion" – 52% in health and 58% in social care activities, together with high levels of "Personal Accomplishment" – 63% of health and 57 % of social personnel.

Taking into consideration the specific socio-economic environment in our country, in which both researched groups are forced to work, it was not surprising that low level of "Reward Dependence" was a necessary prerequisite for the personnel to be employed in health and social care services. The fields of health and social care are permanently underfunded, which makes the income of their employees remarkably low and even some of the lowest possible income even in Bulgaria. However the moral reward for their labor often is symbolic or is missing at all, which is expressed partly in the high pressure and other adverse trends in the psychological climate. In other words, the low "Reward Dependence" is a must for this work, especially in psychiatric hospitals, where any moral or material rewards are sporadic and insufficient. Low score of "Persistence" among social welfare workforce seems to be the comfortable psychological mechanism, effective in the reality of chronic institutional crisis, for coping with increased pressure and oppressive psychological climate. "Persistence" or perseverance, tenacity may have a dual effect

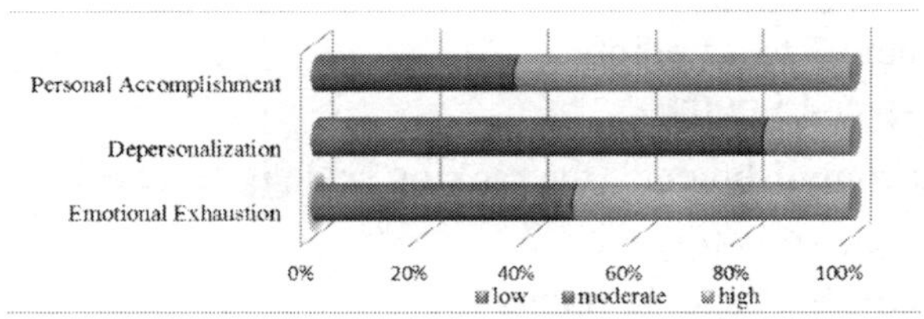

Figure 7.2 Frequency Distribution According to the Levels of "Emotional Exhaustion", "Depersonalization" and "Personal Accomplishment" – Health Care Activities

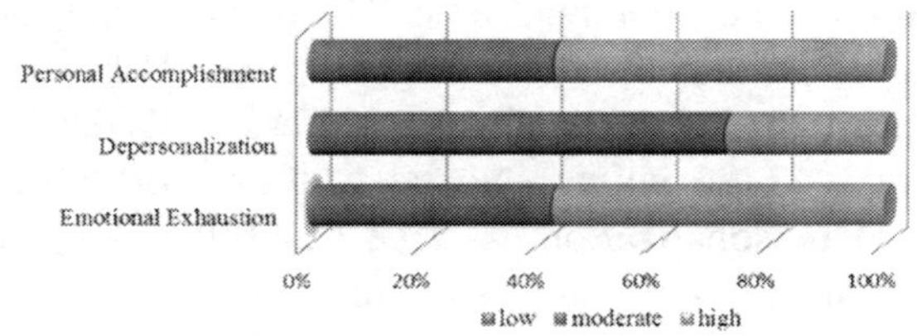

Figure 7.3 Frequency Distribution According to the Levels of "Emotional Exhaustion", "Depersonalization" and "Personal Accomplishment" – Social Care Activities

on the manifestation of burnout. If individuals are persistent and tenacious in work environment that tolerates and encourages that behavior and did not pressed them for specific performance and compliance with regulations, they could function adequately and not burnout. Otherwise, the presence of "Persistence" in an environment that is deeply conservative and limits personal initiative will frustrate and burnout the individual much faster. In our case the observed respondents (health care professionals) were relatively low in "Persistence" as a necessary compromise to their survival in an environment, where each manifestation of perseverance will result in permanent conflict and social marginalization. In other words the personnel in social welfare services were trying to compensate for adverse working conditions with less consistency and reconciliation with the virtual absence of an appropriate reward for their labor. The profile of vulnerability of the workforce employed in the social services was modeled by low "Self-Directedness", indicating their limited potential to achieve goals for self-developing and inability to create new resources. "Self-Directedness" is regarded as the leading and independently acting protective factor, which in our sample was relatively less pronounced, resulting in "Emotional Exhaustion", reduced performance and personal achievements. This adverse phenomenon was partially offset by higher "Cooperativeness", which was an expected trend, considering the integrated team characteristic of the ongoing activities in this field.

On the other hand for health care personnel "Persistence" tends to be high, which means that they show consistency in their activity and thus often conflict with the increased "Pressure" of the environment, which is a dimension of psychological climate. The mixture of high "Persistence" in temperament and high "Pressure" in the psychological climate was a typical illustration of the vulnerability to burnout. The combination of both determinates resulted in disturbingly high levels of "Emotional Exhaustion" and reduced "Personal Accomplishment" for the employees in the researched groups. The scores of "Self-Transcendence" and "Cooperativeness", which were considered protective factors for burnout syndrome, were understated for the health care staff. Therefore they preferred to work individually, did not experience the need of team work and were not able to outgrow their material existence, which was sometimes seriously hampered.

We applied the Spearman's rho correlation coefficient, separately for both groups, to identify, compare and analyze the specific associations between the respondents' personality profile, organizational and psychological climate and the risk of burnout syndrome manifestation (Tables 7.14 and 7.15 – see the Appendix). Due to the similarity of the frequency distributions and the means of the two compared groups, logically, the results repeated the already tested and proven correlations for the whole sample (n = 50). The newly determined associations were due to the specific characteristics of each group, which on the one hand were already analyzed by the mean scores of the scales, and the other by the one-way ANOVA.

Both researched groups were analyzed by one-way ANOVA for associations of group and individual determinants on the level of burnout manifestation (Table 7.16). The analysis demonstrated significance between the scales of TCI-R and IMPC questionnaires and the MBI scales – "Emotional Exhaustion", "Depersonalization" and "Personal Accomplishment". There were five scales that influence the manifestation of burnout syndrome in both samples. In the first group – health care professionals – we observed a bigger dependence of the individual – three vs. two scales of group performance, as opposed to the second group – social welfare workforce – where all associations were due to the group performance.

In the first group "Novelty Seeking" (F=2.91, P=0.042), "Reward Dependence" (F=12.95, P=0.000) and "Cooperativeness " (F=4.21, P=0.011) as characteristics of temperament and character had a significant impact on the "Personal Accomplishment", while the dimensions of group performance as "Pressure" and "Cohesion" were proved to be essential for "Emotional Exhaustion" (F = 7.08, P = .007) and "Depersonalization" (F=2.70, P=0.050). An important point of the analysis was the connection between "Depersonalization" and solidarity of the working environment. After a period of time, the lack of adequate relationships in the workplace could be a prerequisite for health care professional of emotional withdrawal from the needs of patients.

The analysis of the second group showed that five of the eight dimensions of psychosocial climate were modeled burnout syndrome manifestation - "Trust", "Support", "Recognition", "Fairness" and "Innovation". All five listed dimensions were significant for the presence of "Emotional Exhaustion". "Innovation" had a positive impact on the motivation of the respondents and influenced the occurrence of "Depersonalization". "Recognition" is an important motivational element for efficiency and survival of social welfare workforce.

The burn out syndrome influences personal wellbeing and professional performance. In our study we confirmed the contention that healthcare and social workers have personality characteristics that make them more prone to burn out syndrome [5]. Social welfare workforce studies have identified low wages, high caseloads, inadequate training, and poor supervision as key contributors to job burnout and high turnover within this employee population

Table 7.13 One-Way ANOVA for the Effect of Group and Individual Determinants on the Level of Burnout Syndrome by the Type of Activity

	MBI					
TCI-R,	Emotional Exhaustion		Depersonalization		Personal Accomplishment	
	F	P	F	P	F	P
Group I - Healthcare professionals (n=25)						
Novelty Seeking					2.91	0.042
Reward Dependence					12.95	0.000
Cooperativeness					4.21	0.011
Pressure	7.08	0.007				
Cohesion			2.70	0.050		
Group II - Social welfare workforce (n=25)						
Trust	6.82	0.021				
Support	11.02	0.007				
Recognition	6.65	0.023				
Fairness	5.09	0.040				
Innovation	33.05	0.001	2.93	0.041		

[6, 7]. We found that the most important dimension is that of Harm Avoidance, recognized as a factor which leads to the development of burn out. Self-Directedness is assumed to be a protective dimension, as is Cooperativeness. Our results agree closely with the findings of similar studies in the field. The significant relation between personality traits and burn out syndrome are broadly reported in the scientific literature [8–18].

According to our preliminary literature review, not many researchers have used the TCI, particularly not in the field physician and social workers populations, although in recent years their number seems to have increased [19–23]. Mitra and collaborators used the TCI in a comparison of surgeons and anesthesiologists and did not find any significant differences [9]. However, the important link between burn out and personality was demonstrated in numerous publications by using other personality inventories. Traditional research of personality and burn out has focused on the Big Five personality dimensions (extroversion, agreeableness, conscientiousness, neuroticism and openness) [24–29]. Oginska-Bulik has investigated the role of Type D personality in perceiving stress at work and the development of adverse effects of experienced stress, that is, mental health disorders and burn out syndrome [30]. The application of the Eysenck Personality Inventory in an empirical study focused on intensive care healthcare personnel has shown positive correlations of the "external locus of control" with the burn out variables of emotional exhaustion and depersonalization [12]. According to a study in dentistry in the individuals with burn out syndrome, narcissistic and borderline personalities are most frequently found [31]. Few previous studies have investigated and empirically proved that personality affects burn out in professional caregivers by using the NEO Five-Factor Inventory [18, 25, 32].

Other scientific investigations have indicated complex interactions between personality traits and burn out syndrome using various inventories, such as Cattell's 16 Personality Factors Questionnaire, PROSCAN and Scale H of the 16PF [13, 14]. One of these investigations, a study in Sweden, focused on healthcare personnel, has shown that the most important indicators for "belonging to the burn out" were "openness to changes" and "anxiety" and for "belonging to the non-burn out", "emotional stability", "liveliness", "privateness" and "tension" [14]. Furthermore, our findings indicate that the source burn out syndrome may come as much from within individuals, based on their temperamental and character traits, as from outside of them, deriving from the psychological climate dimensions. Healthcare personnel are subject to a variety of work conditions with a potential impact on mental health, inducing job stress, burn out, unspecific psychiatric morbidity or depressive symptoms.

The Psychological climate concept is defined as being a multi-dimensional construct representing shared individual perceptions [2]. Contemporary studies continue to investigate and question the framework of the psychological climate construct, including some projects that identify numerous psychological dimensions [33]. In our study we evaluated specific measures of psychological climate such as Trust, Innovation, Pressure, Recognition, Support, Autonomy, Cohesion and Fairness. Our findings denote that psychological climate might be very important in the development of burn out syndrome. A study conducted in Australia surveyed public hospital employees in the mid-2000s. The results showed that individuals whose perceptions of the organization and environment in which they were working (that is, psychological climate) were more positive, were more likely to appraise changes favorably and report better adjustment in terms of higher job satisfaction, psychological wellbeing (burn out is frequently used as an indicator of reduced wellbeing) and organizational commitment and lower absenteeism and turnover intentions [34]. Investigating possible changes over time in sources of dissatisfaction revealed that factors related to the work environment, rather than individual or demographic factors, were still of most importance to nurses' turnover intentions [35]. The presence of the latter is a direct reflection of depersonalization as a dimension of burn out manifestation. The results of earlier studies suggest that the 3 higher order facets of climate (affective, cognitive and instrumental) affected individual level outcomes of job performance, psychological wellbeing and withdrawal through their impact on organizational commitment and job satisfaction [36].

Burn out appears to be largely dependent on work-related factors [37, 38]. Researchers mostly lock at the negative impact of psychological climate to assess the impact of work on burn out [39]. In addition, work behavior results in various outcomes such as the individual's performance, which is directly related to personality character and temperament. The influence that personality plays in predicting employee behavior has been of interest to organizational psychologists for some time, with particular research emphasis placed on job performance as a criterion variable [40]. Personality determines the way the person interprets events and conceives the interaction between self and the environment [41]. Certain personality traits may enhance the risk of burn out by influencing the individual's response to stressors in the workplace [11]. Commonly, the manifestation of burn out syndrome is associated with and modulated by a variety of personal and environmental factors [42].

In our study, all relationships between personality traits, psychological climate and burn out syndrome, were in the theoretically expected directions. In this research, we focused on one set of individual-level predictors (TCI-R), combined with psychological climate dimensions and show that they are robust predictors of burn out. The correlation results of our study support the new notion that the combination of high Persistence, high Harm Avoidance and weak Character development increase personal vulnerability to the burn out syndrome. In other words, individuals' personalities are a strong predictor of the level of job burnout they experience. But this personality profile is not sufficient to induce burn out on its own. According to our diathesis-stress model, if professional burning is to take place, the vulnerable person develops difficulties in response to a provocative psychological climate. Therefore, the design of our test battery includes the test of Koys and DeCotiis, too [2]. As is shown in the correlation matrices, the correlations between personality and burn out syndrome confirm our theoretical model. The present study identified the main effects of temperament and personality traits on burn out after psychological climate dimensions are taken into account. The causation between the combination of Harm-Avoidance with moderate levels of Pressure and the occurrence of burn out syndrome is observed; that is, generally, a cautious person (high NA), exposed to pressure is more vulnerable to Emotional Exhaustion. The lack of Personal Accomplishment is positively and critically related with High Persistence combined with Low Autonomy, that is, an individual with High Persistence positioned in a working environment, characterized by low autonomy and innovation, is at risk of Personal Accomplishments decline and *vice versa*. High Cooperativeness combined with High Cohesion is in negative association with Depersonalization. Depersonalization itself would appear to be unlikely in healthcare employees considering that the specificity of their professional duties is in direct opposition to distancing oneself from patients and co-workers. Our findings demonstrate strong relationships among the combination of Self-Directedness with Autonomy and 2 of the burn out subscales – Personal Accomplishments and Emotional Exhaustion. The directions of the associations are affected by the combination of the scales' values for Self-Directedness with Autonomy. Specifically, the relationships appear as follows – High Self-Directedness coupled with High Autonomy is positively associated to Personal Accomplishment and negatively to Emotional Exhaustion. In this causation we recognize a potential protective factor. The results confirm the theoretical assumption that personal vulnerability interacts reciprocally with the psychological climate of the organization to induce the burn out syndrome. Thus, they create a vicious cycle of cause and effect interaction. The findings of this study highlight the importance of becoming aware that personality traits, psychological climate and burn out are a triune multi-faceted phenomenon. Given that our findings have shown significant correlation between scales' liaisons, they create the basis for further discussion and research.

Using personality and psychological climate theories as conceptual frameworks, we hypothesized that the type of burn out would be specifically related with the internal associations between personality dimensions and psychological climate determinants. Our results were supportive and, in addition, a number of notable findings emerged. The present study makes a noteworthy contribution to the field by demonstrating the magnitude of individual differences, particularly personality dimensions in combination with psychological climate settings as moderators of burn out syndrome manifestation. The present research provides empirical evidence in support of the significant relations between the scales and, more specifically, the internal coherence among their subscales. Notwithstanding the small sample size (low statistical power), the sufficient occupational heterogeneity gave us the ability to generalize the findings of the pilot study. We expect to substantiate the present results in our ongoing major study that covers a larger area and processes a bigger sample size.

This study is unique in that it includes a three dimensional construct of personal vulnerability, anomalies of the psychological climate and burn out. Knowledge of predisposing factors can be useful for the prevention of emotional exhaustion and depersonalization within a work organization [43]. We anticipate that identifying these causal phenomena early, rather than focusing on later occurring symptoms, will enable the introduction of adequate schemes aimed at the prevention of professional burn out among healthcare specialists. Unfortunately, burn out is difficult to prevent. However, the need for preventive programs aimed at reducing stress experienced at work and protecting the health of medical personnel, was strongly suggested by the results obtained in our research. The programs should be implemented in to organizations with the aim of changing the system of rewards, to improve social relations between both employees and employers and employees and co-workers and to reduce workload and the sense of threat [31]. It is necessary to emphasize team work and to provide positive psychological feedback which is necessary for the working person in order to achieve work satisfaction. It is frequently necessary to change positions and duties so as to maintain the person's interest in his/her work. However, this shift should not be too frequent and it should respect the qualifications of the person (i.e., it is wrong to *de facto* downgrade someone just in order to change his position). Also, the early recognition of depression development and of co-morbidity is essential, because they lead to early and fuller treatment [44]. The specialized natural response to the enormous social damage due to burn out is the development of strategies and behavioral models aimed at reducing the production of stress and to overcome it. The cardinal approaches have two aspects: organizational (including developing and implementing programs to assist employees in overcoming stress, training in stress management, using various intervention components to reduce stress) and individual (focused primarily on the detection and removal of individual reasons leading to stress). It is believed that both approaches are positively affected by public support. It is essential that employees are aware of the health effects of occupational stress and burn out, so that they develop skills to overcome them and to restore their own energy reserves. Corporate programs prepare employees to deal with the negative effects of everyday stress, to regulate their emotional state and to improve their competence in the field of interpersonal interactions [45, 46]. According to a study conducted among US physicians, the best prevention for burn out is to promote personal and professional wellbeing on all levels: physical, emotional, psychological and spiritual. This must occur throughout the professional life cycle of physicians, from medical school through to retirement. It is a challenge not only for individual physicians in their own lives, but also for the profession of medicine and the organizations in which physicians work [11].

References

[1] Cloninger, C.R., Svrakic, D.M. & Przybeck, T.R. (1993). A psychobiological model of temperament and character. Archives of General Psychiatry 50, 975–990.]

[2] Koys, D.J. & DeCotiis, T.A. (2001). Inductive measures of psychological climate. Human Relations 44, 265–285.

[3] Maslach, C., Jackson, S.C. & Leiter, M.P. (1996). Maslach Burnout Inventory. Palo Alto, California: Consulting Psychologists Press

[4] Cloninger, C.R., Przybeck, T.R. & Svrakic, D.M. (1991). The Tridimensional Personality Questionnaire: U.S. normative data. Psychological Reports 69 (3) 1047–1057.

[5] Gundersen, L. (2001). Physician burnout. Annals of Internal Medicine 4, 145–148.

[6] Annie E. Casey Foundation. The unsolved challenge of system reform: The condition of the frontline human services workforce (2003) Retrieved from AECF website: http://www.aecf.org/upload/publication files/the%20unsolved%20challenge.pdf Feb 2013

[7] General Accounting Office Child welfare: HHS could play a greater role in helping child welfare agencies recruit and retain staff (Publication No. GAO-03-357) General Accounting Office, Washington, DC (2003).

[8] Kumar, S. (2007) Burnout in psychiatrists. World Psychiatry 6, 186–189.

[9] Mitra, S., Sinha, P.K., Gombar, K.K. & Basu, D. (2003). Comparison of temperament and character profiles of anesthesiologists and surgeons: a preliminary study. Indian Journal of Medical Sciences 6, 431–436.

[10] Piedmont, R.L. (1993). A Longitudinal Analysis of Burnout in the Health Care Setting: The Role of Personal Dispositions. Journal of Personality Assessment 61 (3) 457–473.

[11] Spickard, A., Gabbe, S.G. & Christensen, J.F. (2002). Mid-Career Burnout in Generalist and Specialist Physicians. Journal of the American Medical Association 288 (12) 1447–1450.

[12] Buhler, K.E. & Land, T. (2003). Burnout and personality in intensive care: An empirical study. Hospital Topics: Research and Perspectives on Health Care 8, 5–12.

[13] Eastburg, M.C., Williamson, M., Gorsuch, R. & Ridley, C. (1994). Social Support, Personality, and Burnout in Nurses. Journal of Applied Social Psychology 24 (14) 1233–1250.

[14] Gustafsson, G., Persson, B., Eriksson, S., Norberg, A. & Strandberg, G. (2009). Personality traits among burnt out and non-burnt out health-care personnel at the same workplaces: a pilot study. International Journal of Mental Health Nursing 18, 336–348.

[15] Deckard, G., Meterko, M. & Field, D. (1994). Physician Burnout: An Examination of Personal, Professional, and Organizational Relationships. Medical Care 32 (7) 745–754.

[16] Glasberg, A.L., Eriksson, S. & Norberg, A. (2007). Burnout and "stress of conscience" among healthcare personnel. Journal of Advanced Nursing 13, 392–403.

[17] Keinan, G. & Melamed, S. (1987). Personality characteristics and proneness to burnout: A study among internists. Stress Medicine 3 (4) 307–315.

[18] Narumoto, J., Nakamura, K., Kitabayashi, Y., Shibata, K., Nakamae, T. & Fukui, K. (2008). Relationships among burnout, coping

style and personality: Study of Japanese professional caregivers for elderly. Psychiatry and Clinical Neurosciences 62, 174–176.

[19] Pejuskovic, B. & Lecic-Tosevski, D. (2011). Burnout in psychiatrists, general practitioners and surgeons. World Psychiatry 10, 78.

[20] Gárriz, M. & Gutiérrez, F. (2009). Personality disorder screenings: a meta-analysis. Actas Espanolas de Psiquiatria 37 (3) 148–152.

[21] Goekoop, J.G., DeWinter, R.F.P. & Goekoop, R. (2011). An Increase of the Character Function of Self-Directedness Is Centrally Involved in Symptom Reduction during Remission from Major Depression. Depression Research and Treatment In press.

[22] Juradoa, D., Gurpequi, M., Moreno, O., Fernández, M.C., Luna, J.D. & Gálvez, R. (2005). Association of personality and work conditions with depressive symptoms. European Psychiatry 20 (3) 213–222.

[23] Naito, M., Kijima, N. & Kitamura, T. (2000). Temperament and Character Inventory (TCI) as Predictors of Depression among Japanese College Students. Journal of Clinical Psychology 56 (12) 1579–1585.

[24] Bakker, A.B., Van Der Zee, K.I., Lewig, K.A. & Dollard, M.F. (2006). The relationship between the Big Five personality factors and burnout: a study among volunteer counselors. Journal of Social Psychology 146, 31–50.

[25] Gandoy-Cregoa, M., Clementeb, M., Mayán-Santosa, J.M. & Espinosab, P. (2009). Personal determinants of burnout in nursing staff at geriatric centers. Archives of Gerontology and Geriatrics 48 (2) 246–249.

[26] Barrick, M.R. & Mount, M.K. (1991). The Big Five personality dimensions and job performance: a meta-analysis. Personnel Psychology 44, 1–26.

[27] Hurtz, G.M. & Donovan, J.J. (2000). Personality and job performance: the Big Five revisited. Journal of Applied Psychology 85, 869–879.

[28] Krasner, M.S., Epstein, R.M., Beckman, H., Suchman, A.L., Chapman, B., Mooney, C.J. & Quill, T.E. (2009). Association of an Educational Program in Mindful Communication With Burnout, Empathy, and Attitudes Among Primary Care Physicians. Journal of the American Medical Association 302 (12) 1284–1293.

[29] Kim, H.J., Shin, K.H. & Swanger, N. (2009). Burnout and engagement: A comparative analysis using the Big Five personality dimensions. International Journal of Hospitality Management 28, 96–104.

[30] Oginska-Bulik, N. (2006). Occupational stress and its consequences in healthcare professionals: the role of type D personality. International Journal of Occupational and Environmental Health 10, 113–122.

[31] Alemany, M.A., Berini, A.L. & Gay, E.C. (2008). The burnout syndrome and associated personality disturbances. The study in three graduate programs in Dentistry at the University of Barcelona. Medicina Oral Patologia Oral Cirugia y Bucal 13, 444–450.

[32] Schimpf, T. (2009). Personality traits and burnout in clinical psychologists. Dissertation 2009; 112 p http://gradworks.umi.com/33/55/3355056.html accessed 07.06.2012

[33] Patterson, M.G., West, M.A., Shackleton, V.J., Dawson, J.F., Lanthorm, R., Maitlis, S., Robinson, D.L. & Wallace, A.M. (2005). Validating the organizational climate: Links to managerial practices, productivity and innovation. Journal of Organizational Behavior 26 (4) 379–408.

[34] Martin, A.J., Jones, E.S. & Callanc, V.J. (2005). The role of psychological climate in facilitating employee adjustment during organizational change. European Journal of Work and Organizational Psychology 14 (3) 263–289.

[35] Coomber, B. & Barriball, K.L. (2007). Impact of job satisfaction components on intent to leave and turnover for hospital-based nurses: a review of the research literature. International Journal of Nursing Studies 44 (2) 297–314.

[36] Carr, J.Z., Schmidt, A.M., Ford, J.K. & DeShon, R.P. (2003). Climate perceptions matter: A meta-analytic path analysis relating molar climate, cognitive and affective states, and individual level work outcomes. Journal of Applied Psychology 88 (4) 605–619.

[37] Crabbe, J.M., Bowley, D.M., Boffard, K.D., Alexander, D.A. & Klein, S. (2004). Are health professionals getting caught in the crossfire? The personal implications of caring for trauma victims. Emergency Medicine Journal 21 (5) 568–572.

[38] Chiriboga, D.A. & Bailey, J. (1986). Stress and burnout among critical care and medical surgical nurses: a comparative study. Critical Care Quarterly 9 (3) 84–92.

[39] García-Izquierdo, M. & Ríos-Rísquez, M.I. (2012). The relationship between psychosocial job stress and burnout in emergency department: An exploratory study. Nursing Outlook. Article in press. http://ac.els-cdn.com/S0029655412000395/1-s2.0-S0029655412000395-main.pdf?_tid=4aeb22033ddde00ec12964e3bb73e3e1&acdnat=1339409481_c3fd885554fc9e53c044163d907f1906 Accessed 11.06.2012.

[40] Byrne, Z.S., Jason, S., Kenneth, T. & Hochwarter, W. (2005). The interactive effects of conscientiousness, work effort, and psychological climate on job performance. Journal of Vocational Behavior 66 (2) 326–338.

[41] Cooper, C.L. & Baglioni, A.J. (1988). A structural model approach toward the development of a theory of the link between stress and mental health. British Journal of Medical Psychology 61, 87–102.

[42] Paisley, K. & Powell, G.M. (2007). Staff burn-out prevention and stress management. Child and Adolescent Psychiatric Clinics of North America 16 (4) 829–841.

[43] Giliberta, D. & Dalozb, L. (2008). Disorders associated with burnout and causal attributions of stress among health care professionals in psychiatry. European Review of Applied Psychology 58 (4) 263–274.

[44] Iacovides, A., Fountoulakis, K.N., Kaprinis, S. & Kaprinis, G. (2003). The relationship between job stress, burnout and clinical depression. Journal of Affective Disorders 75 (3) 209–221.

[45] Pejuskovic, B., Lecic-Tosevski, D., Priebe, S. & Toskovic O. (2011). Burn-out syndrome among physicians – the role of personality dimensions and coping strategies. Psychiatria Danubina 23, (4) 389–395.

[46] Plana, A.B., Fabregat, A.A. & Gassió, J.B. (2003). Burnout syndrome and coping strategies: a structural relations model, Psychology in Spain 7 (1) 46–55.

8

Regression and Structural Equation Modeling of Burnout Components in Health Care

Sandor Rozsa PhD, Assistant Professor,
Eotvos Lorand University, Psychological Institute,
Department of Personality and Health Psychology; Washington University School of Medicine, St. Louis, Department of Psychiatry, psychologist

Donka Dimitrova PhD, Engineer, Associate Professor of Health Care Management, Statistical Advisor,
Department of Health management, Health Economics and General Practice, Faculty of Public Health, MUP

The application of statistical methods in psychology originated in the late 19th and early 20th century. Referring to the depth, complexity and possibilities for large datasets processing that implies methods for evaluation of the properties of the tools, as well as the attempts to determine relationships and building formal causal models using regression analysis and structural models with the introduction of latent variables.[1,2]

In the previous chapters we discussed the validity and reliability assessment of the test battery comprising the Bulgarian versions of TCI-R, IMPC and MBI. Our results revealed that Cronbach's alpha coefficients of the scales of three instruments were all satisfactory and with few exceptions most alphas exceeded 0,70. There was no significant difference between males' and females' scores of TCI-R, IMPC, MBI and the scales did not correlate highly with age or work experience. (Table 8.1)

Correlations analyses between burnout components and study variables proved that Reward dependence (RD), Persistence (PS) and Cooperativeness (CO) positively related to Personal accomplishment (PA) while positive relationships also existed between Harm avoidance (HA) and two burnout dimensions (EE and DP).

Drozdstoy St. Stoyanov (Ed.), New Model of Burn Out Syndrome: Towards Early Diagnosis and Prevention, 107–116.

Table 8.1 Descriptive statistics of the TCI-R, IMPC and MBI

Scales	No. items	Cronbach's α	Males		Females		t	p	Correlation with age	Correlation with work experience
			Mean	*SD*	*Mean*	*SD*				
TCI-R										
Novelty seeking (NS)	35	,78	106,78	13,28	106,69	11,93	,06	,95	,08	,10
Harm avoidance (HA)	33	,82	95,67	13,54	97,44	12,49	1,10	,27	,16**	,18**
Reward dependence (RD)	30	,68	98,66	11,51	99,13	9,75	,37	,71	,16**	,17**
Persistence (PS)	35	,87	120,30	16,47	121,44	15,80	,57	,56	,05	,10
Selfdirectedness (SD)	40	,86	125,66	16,89	123,76	17,87	,86	,38	,14*	,13*
Cooperativeness (CO)	36	,74	116,73	12,29	118,47	12,83	1,10	,27	,16**	,18**
Self-transcendence (ST)	26	,82	79,30	13,65	81,38	12,27	1,30	,19	,17**	,19**
IMPC										
Autonomy	5	,87	19,43	3,92	18,88	3,97	1,12	,26	,17**	,20**
Cohesion	5	,89	17,59	4,57	17,93	4,31	,62	,53	,21**	,24**
Trust	5	,92	20,24	4,47	19,74	4,63	,87	,38	,05	,09
Pressure	5	,47	16,11	3,52	15,82	3,56	,65	,51	,07	,09
Support	5	,92	19,97	4,48	19,27	4,79	1,19	,23	,06	,11
Recognition	5	,60	17,54	3,45	17,20	3,46	,79	,43	,07	,12*
Fairness	5	,89	19,29	4,18	19,13	4,45	,29	,76	,10	,13*
Innovation	5	,93	18,89	4,68	19,10	4,53	,35	,72	,07	,11
MBI										
Emotional exhaustion (EE)	9	,91	21,19	11,87	22,11	11,19	,64	,51	,12*	,12*
Depersonalization (DP)	4	,63	7,31	5,01	6,61	4,32	1,24	,21	,06	,05
Personal accomplishment (PA)	8	,61	32,25	6,09	31,71	5,78	,73	,46	-,08	-,07

**p<,01; *p<,05

To test further the initial hypothesis of the conceptual model of the Project we performed regression analysis of the field study data.

8.1 Hierarchical Stepwise Regression Analysis Predicting Burnout Components

Stepwise regression analysis provides models omitting non-significant predictors. Considering the adopted conceptual model hierarchical stepwise regression analyses was used to test the *direct effects* of *sociodemographic variables* (sex, age, work experience), *psychological climate scales* (IMPC) and *personality traits* (TCI-R) on three burnout components (dependent variables EE, DP, PA).

In the first step of the hierarchical regression procedure, the demographic predictor variables: age, gender, and work experience were block entered providing the variance accounted for in this group of independent variables. In the second step psychological climate scales (IMPC) were entered into the step 1 model. Next, we repeated the entry of the first and second step and block entered the personality traits (TCI-R) predictor variables.

Since there were no significant predictors from sociodemographic variables, Table 8.2 only presents steps where the significant predictors were found.

The results indicated that the variance accounted for (R^2) with the first IMPC predictors equaled ,26, which was significantly different from zero. Next, TCI-R dimension scores were entered into the regression equation. The change in variance accounted for (Δ R2) was equal to ,02 for EE (F/3, 294/= 40,11, p<,001) and to ,09 for PA (F/5, 292/= 15,24, p<,001), which were statistically significant increases in variances accounted for over the step one models.

The obtained models describing the impact of significant predictors on the burnout components explained about 30% of the responses received for Emotional exhaustion and over 20% for Personal accomplishment and in psychology models with such predictive power are considered significant by numerous authors.[4–6]

Table 8.2 Hierarchical regression analyses evaluating predictors of burnout components

Dependent variable	Independent variables		β	R^2	ΔR^2
Emotional exhaustion (EE)	Step 1:	IMPC		,26**	
		Pressure	,48**		
		Support	-,26**		
	Step 2:	TCI-R		,28**	,02**
		Harm avoidance	,16**		
Model: F(3, 294)= 40,11, p<,001					
Depersonalization (DP)	Step 1:	IMPC		,14**	
		Pressure	,32**		
		Trust	-,26**		
Model: F(2, 295)= 24,62, p<,001					
Personal accomplishment (PA)	Step 1:	IMPC		,15**	
		Autonomy	,20**		
		Pressure	-,17**		
		Cohesion	,12*		
	Step 2:	TCI-R		,24**	,09**
		Persistence	,36**		
		Self-transcendence	-,13*		
Model: F(5, 292)= 15,24, p<,001					

Note. Betas reported are those from the step at which the variable was significant and entered into the equation.
**p<.01; *p<.05.

Our analysis confirmed to a great extent the hypotheses of relationship between organizational factors and personal characteristics in terms of vulnerability to burnout. Our study, as well as numerous other studies gave us reason to assume that temperament and character traits (TCI-R scales) have both direct and indirect effect on burnout dimensions. However the regression analyses based on the additive model of predictors did not provide insight of the indirect mediation effects.

Studying complex relationships usually requires other types of analysis in order to assess the role of the actual characteristics of the psychological attitude as mediators and since our study was somehow limited by the sample size, the complexity of the conceptual model and the many variables in it we felt further modeling was needed using concepts from similar studies.[4–6]

8.2 Structural Models of Burnout Components

To test the meditational effects of psychological climates, structural modeling (Structural Equation Modeling) was performed using the main predictors from our previous regression analyses as this approach works in confirmatory mode.

We included as predictors and mediators the main personality and organizational variables identified by other authors - HA, PS, RD, Pressure, Support and Fairness, which largely coincided with the above models. In the process of creating the structural model we also considered adding predictors that were near the significance level in our regression analyses (PS).[6] (Figure 8.1)

Referring to the IMPC scales Koys and DeCotiis proposed descriptive dimensions[7]:

1. **Norms and organisational structure** (Cohesion, Trust, Support, Fairness);
2. **Reward and control mechanisms** (Autonomy, Pressure, Recognition, Innovation).

Structural Equation Modeling also allowed creating latent variables, so in the psychological climate we defined two respective latent variables. Our analysis partially supported the two dimensions, as Reward and control mechanisms consists only two scales (Autonomy and Pressure). The models, although not directly involving the dimensions of organizational culture such as leadership style, management decisions taking modes, organizational adaptability, etc. complied with the concept, confirming the effect of Autonomy and Pressure. (Figure 8.2)

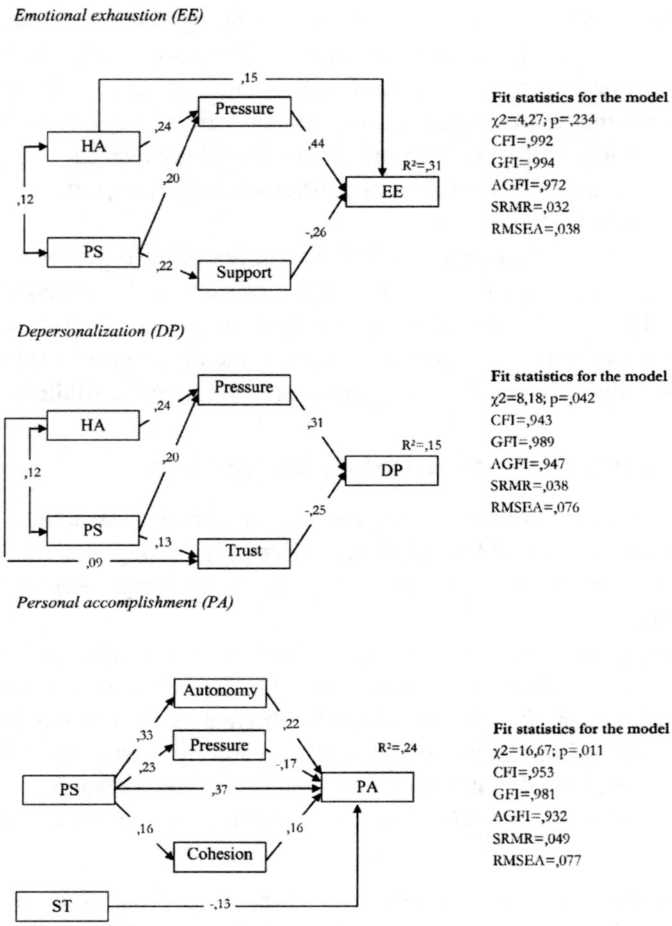

Figure 8.1 Structural Models of Burnout Components

The introduction of latent organizational and cultural factors significantly increased the predictive power of the models to 56% in EE and 30% in DP. These results suggested that the levels of burnout were determined to a greater extent by factors of the organizational culture than by personality traits which were the focus of research in our study.

The effective management and prevention of burnout syndrome rest in the field of management decisions and policies of healthcare organizations. To maximize efficiency of management practices it is essential to integrate psychological profiling and organizational culture assessments.

Emotional exhaustion (EE)

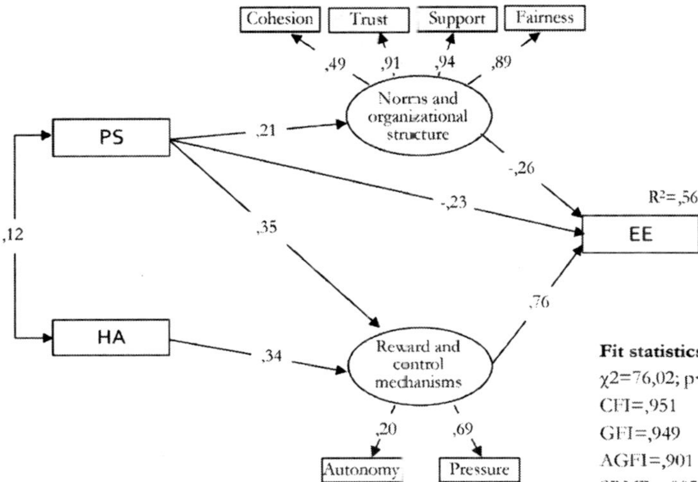

Fit statistics for the model
χ2=76,02; p<,001
CFI=,951
GFI=,949
AGFI=,901
SRMR=,085
RMSEA=,088

Depersonalization (DP)

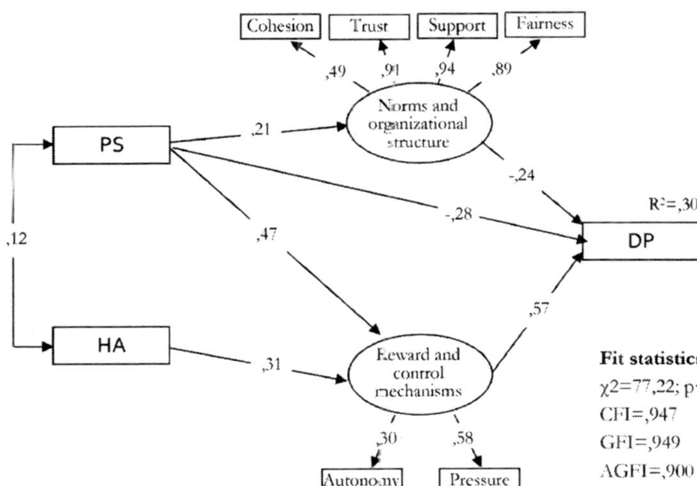

Fit statistics for the model
χ2=77,22; p<,001
CFI=,947
GFI=,949
AGFI=,900
SRMR=,078
RMSEA=,089

Personal accomplishment (PA)

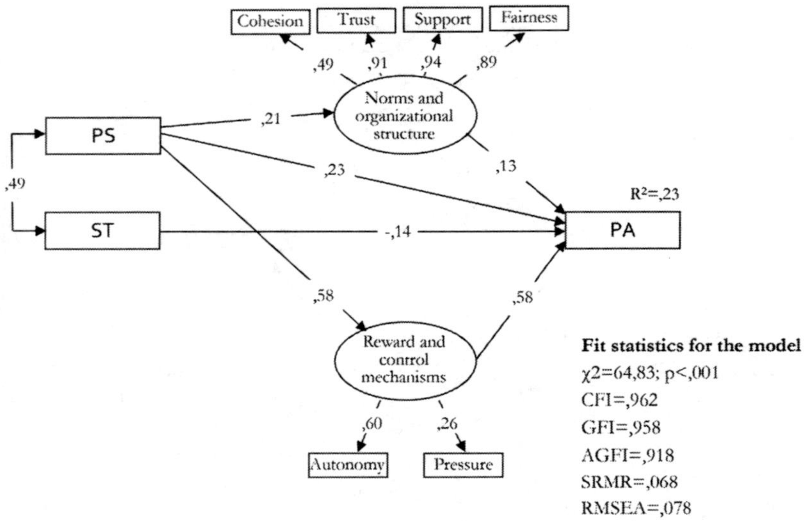

Figure 8.2 Structural Models of Burnout Components with Introduced Latent Variables

References

[1] Kalinov K. Statisticheski metodi v povedencheskite I socialni nauki. (Statistical methods in behavioral and social sciences. in Bulgarian). NBU, Sofia, 2001

[2] Kalinov K. Psichologia i statistika: perspektivite na 21 vek. (Psychology and statistics: 21st century perspectives. in Bulgarian). III National congress in psychology, Sofia, 28–30 October 2005 http://bjop.files.wordpress.com/2008/10/krasimir-kalinov-doklad.pdf

[3] B.W. Swider, R.D. Zimmerman Born to burnout: A meta-analytic path model of personality, job burnout, and work outcomes Journal of Vocational Behavior 76 (2010) 487–506

[4] H.J. Kim et al Hotel job burnout: The role of personality characteristics. Hospitality Management 26 (2007) 421–434

[5] À.B. Bakker, K. I. Van Der Zee, K. A. Lewi, M. F. Dollard. The Relationship Between the Big Five Personality Factors and Burnout: A Study Among Volunteer Counselors. The Journal of Social Psychology, 2006, 146(1), 31–50.

[6] M. Serec, B. Bajec, D. Petek, I. Švab, P. Seliè. A structural model of burnout syndrome, coping behavior and personality traits in professional soldiers of the Slovene armed forces. Zdrav Vestn 2012; 81: 326–36

[7] Koys, D. J., DeCotiis, T. A. Inductive measures of psychological climate. Human Relations, (1991) 44(3), 265–285.

9

Anxiety and Depression as State Predictors for Burn Out in Health Care

Slavka Toshkova-Hristozova PhD,
Department of Language and Specialized Training, MUP

Drozdstoy Stoyanov MD, PhD, Full Professor of Psychiatry,
Medical Psychology and Person Centered Medicine,
Faculty of Medicine, MUP,
Vice Chair Executive Committee PSIG,
Royal College of Psychiatrists, Visiting Fellow,
University of Pittsburgh

Kaloyan Haralampiev PhD, Associate Professor,
Department of Sociology, Faculty of Philosophy,
University of Sofia "St. Clement of Ohrid"

Anxiety is an emotional situational reaction to different types of stressors (Hanin, 1978). In its nature it is a negative diffuse emotion, arousing emotional tension and experience of failures. The mistakes made are explained through character traits which the personality perceives as personal faults. Anxiety results are low self-esteem, lack of confidence, self-humiliation and other unpleasant experiences predisposing to depressive mood.

Anxiety is perceived as first stage of stress which is 'nonspecific body response to every requirement presented to it' (Selye, 1982). Stress is an adaptive reaction of the organism in which it has to find internal reserves to cope with the situation. Reacting in a stressful situation or under the pressure to a 'stressor' is always in the form of emotional experience. Thus anxiety is a reaction loaded with negative consequences for the personality and affects its emotional resources. It consumes its adaptive energy and leads to emotional exhaustion- the final stage of stress.

According to the theory of emotions (James, 1890) the individual emotional behavior is not caused by feelings but is a result from it. In this sense

Drozdstoy St. Stoyanov (Ed.), New Model of Burn Out Syndrome: Towards Early Diagnosis and Prevention, 115–130.

the theory for self-perception deals with the causes for emotional behavior (Bem,1972; Laird, 1974, 1984, Georgieva, 2012) which may express anxiety or depression.

As far as the emotional resource of the personality is an individual peculiarity, then the frequency of the emotional response to irritations from micro-stressors at the work place, as well as their duration, affect health. Provoking anxiety in the work process transforms psycho emotional factors at the work place into an object for health risk evaluation (Tsenova, 2002; Tsenova, 1993). Occupational safety has been presented as a criterion for the culture of the healthcare institution (Flin, 2006).

Anxiety is among the indicators for individual health (Ivanova, 2003). The impact, exerted by psychosocial stressors on the labour environment (Ilieva, 2007), the dependence between increased labour activity, anxiety and emotional exhaustion in teachers (Hristov, 2010), the development of metabolic disorders, frustration among physicians and teachers (Tulevski, 1986; Hristov, 2003; Asenova, 2003; Nunev, 1999 etc.), are among the contemporary problems focusing on the mental health of the people in working age.

Anxiety is a determinant for the level of work organization and the conditions in which the labour process is performed (Hristov (2009). The anxiety state is proved by empathy (Zhivkova, Hristina 1996). If in the 'subject-subject' professions empathy and chronic stress are basic psychosocial factors leading to anxiety and predisposing to emotional exhaustion (Hristov, 2004), then a micro stressor of crucial importance in all employees is the psycho climate on the work place (Tsenova, 1998; Tilov, 2012). Clarifying the nature of anxiety among people in working age infers the generalized conclusion that practicing a profession with scope of activity relationships of the 'human- human' type contains an increased risk of emotional exhaustion than practicing professions with other scope of activity.

With a view to the dependence between the scope of activity and health, defining specific psychoemotional stressors, causing anxiety in the workplace to healthcare professionals would not be complete if those specific stressors, which affect the part of doctors, engaged in medical student training, are ignored. The clinical environment in the university hospital is the base where they work, teach and convey their knowledge and skills. Apart from an environment for acquiring theoretical knowledge and skills, it is an environment in which professionals teach and do research. The setting of the university clinics is a working environment for medical and non medical staff and a place where hospital, educational and academic environments are united. Thus the complex of working environment conditions is a cause for deposition of impact on the side of stressors which in other conditions act independently. Combining influences increases the intensity, strength of emotional response with which the working healthcare professionals react. The integrated action of stressors in the working environment of the university hospitals in practice enriches the means for exerting emotional pressure on the physician and through it – also for provoking anxiety and manifestation of depressive state.

Placing the lecturer in an environment with multiple layer influences is a specificity resulting from the integrated type of work which he/she performs. The peculiarities of labor and configuration of contexts in implementing practical activity, transform the peculiarities of the working activity in psychosocial stressors (Figure 9.1)

The assumption that combining activities and work in combined working conditions leads to excess work activity and physical and mental overload, is

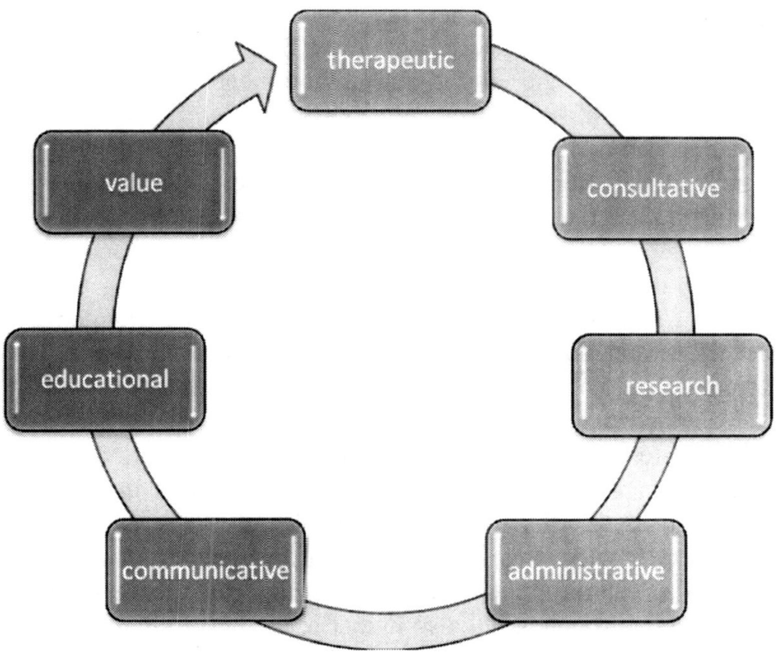

Figure 9.1 Specific features of the working environment at the University hospital

due to the combination of functions and assumption of additional responsibilities. On the one hand, they may be interpreted as vertical workload and on the other, they are assessed as stimulating self actualization and motivating academic development. I.e. apart from the purely objective prerequisites for anxiety occurrence, subjective, arising from the individual choice are also present. According to the five-factor personality model (William McDougal, 1932) the personality dimensions which motivate the subjective labour motivation are extroversion, emotional stability, agreeableness, conscientiousness and openness. Special attention is paid to the element emotional stability which in this case is provoked by different variations of combined activity, according to the functions and aims performed at the specific moment.

Considering the combination of medical and educational environment, studying anxiety among lecturers from the university hospitals presupposes accounting for the specificities of both working environments. In the setting of every healthcare institution micro social stressors causing anxiety operate – resource provision of activities, complying with the requirements of the patients as consumers of healthcare services, psycho climate, labour remuneration, professional qualification etc. (Borisov, 2000). To them we add those operating in the educational environment of medicine and the academic environment in medical university- competitive start to occupy a post, writing and defending PhD thesis, scientific publications, appraisal, writing educational literature etc. It is assumed that, with a view to the professional qualification of the physician, his anxiety is influenced in a high degree by stressors operating in the educational environment and connected to the teaching of students and performing teaching functions.

The strife for demonstration of skills at work and achieving good results correlates with increased risk from depression and burn-out incidence (Hristov, 2009; Ivanova, 1992). In general, burn out is a condition provoked by the nature of labour, work place stress and lack of satisfaction. Psychosomatic complaints, professional burnout, depression, life and behavior satisfaction related to health, are among the indicators for stress in the educational environment (Tsenova, 1996), incl. that in medicine.

The presence of wide range of psychosocial stressors provoking emotional resistance of the lecturer in medicine increases the level of stress not only in the educational environment but also in that of the university hospital. With a view to the fact that in stress the anxiety stage is followed by the stages of resistance and emotional exhaustion, then the study of the manifestations of anxiety aims to undertake timely measures to react to and limit the negative factors- potential causers of burnout. The anxiety evaluation in combined working conditions aims also at 'capturing' and registering indicators for negative impact influencing the mental health of the physician. In this sense, study of anxiety in a combined working environment in medicine contains four contexts- health of the healthcare professionals, the quality of their treatment activity, and the quality of education in medicine and management of an environment with complex nature. The complexity of the problem assumes solving many research problems but three of them have basic significance:

- To diagnose the working environment through anxiety assessment among the employees
- To define the psychosocial stressors specific for the environment which lead to anxiety
- To estimate the tendency for burn our occurrence among lecturers in medicine giving an account for the level of anxiety and the individual peculiarities of the personality.

It is assumed that anxiety and depressive symptoms among lecturers in the university hospitals describe an image which presents initial stage of professional burnout, known as 'flame out'. It is accepted that the syndrome known as 'rust out' still does not exist in its complete state.

Recording anxiety among lecturers of medicine is based on a survey encompassing all medical universities in Bulgaria (2012). 115 people participate in it and they are distributed in three groups- physicians teaching medicine and working at a university hospital (complex working environment), physicians who have only treatment activity in a university hospital or work as general practitioners and lecturers whose professional realization is only in the educational environment of the medical university.

The methods of anxiety study use a specialized test (neurotic depressive test of T. Tashev, 1976). The term 'neuroticism', used in the test, is equal to anxiety according to the contemporary classification. Anxiety is a prerequisite for developing burn out syndrome. Indirectly anxiety in the labor environment may be assessed though job satisfaction.

The average age of the participants in the survey is 48,69 with standard deviation ± 10,33 CI 95% (Figure 9.2).

The distribution according to sex reveals predominance of women (58,26 %). Men make up a group of 41,74 %.

Apart from the demographic characteristics sex and age the anxiety survey among lecturers is influenced by the factor years of employment. The highest relative share comprises the group of lecturers with length of service above 20 years - 30,31% (n = 115). The share of physicians with practice between

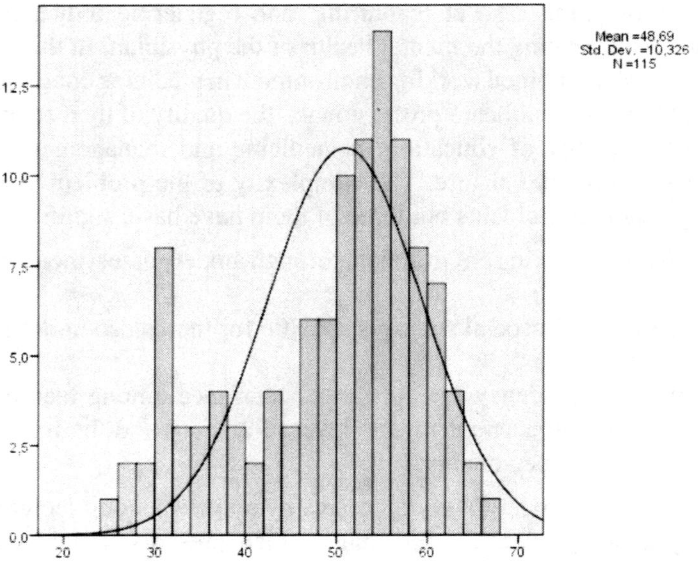

Figure 9.2 Histogram of the average age

25–30 years in the system of the medcial education is 26,96%. Those who have worked up to 15 years are 22,61 %, and those working up to 5 years, the relative share is 26,96%.

Habilitation staff participating in the survey are 23,85 % (Figure 9.3).

According to the analysis of the test results, the share of the practically healthy is 35,65%. Those in which a pre neurotic state is registered are 28,70%.. The group with marked neuroticism encompasses 24,35% of participants (n = 115). The internal distribution reveals that the affected in a mild degree are 5,22%. Those in which neuroticism is in an average degree are 13,91%. In 5,22% the condition is in a severe degree. I.e. there is an overlapping in the degrees 'mild' and 'severe'(Table 9.1).

With all lecturers- from the complex working environment and the 'classical 'working environment of the medical university- the number of practically healthy participants predominates. In comparison to them the group of physicians, rendering medical aid, a match in the practically healthy and neurotics in 'average degree' (23,80%) is recorded. In this group the highest relative share is with the physicians with pre neurotic state (33,30%). In both groups lecturers these who are in the pre neurotic state, comprise the second largest groups- 16,70% for employees of the university hospital and 32,80%

Table 9.1 Clinical-Diagnostic Picture According to the Nature of the Practical Activity (n=115)

Clinical status	Physicians with teaching activity			Lecturers			Clinicians			Total		
	num	Ð[%]	Sp	num	P[%]	Sp	num	P[%]	Sp	num	P[%]	Sp
Practically healthy	13	43,30	13,74	23	35,9	10,00	5	23,80	19,04	41	35,65	7,48
Pre neurotic conditions	5	16,70	16,68	21	32,8	10,24	7	33,30	17,81	33	28,70	7,87
Neuroticism in mild degree	1	3,30	17,94	3	4,7	12,22	2	9,50	20,73	6	5,22	9,08
Neuroticism in average degree	4	13,30	16,98	7	10,9	11,78	5	23,80	19,04	16	13,91	8,65
Neuroticism in severe degree	2	6,70	17,68	3	4,7	12,22	1	4,80	21,38	6	5,22	9,08
No answer	5	16,70	16,68	7	10,9	11,78	1	4,80	21,38	13	11,3	8,78
Total	30	100		64	100		21	100		115	100	

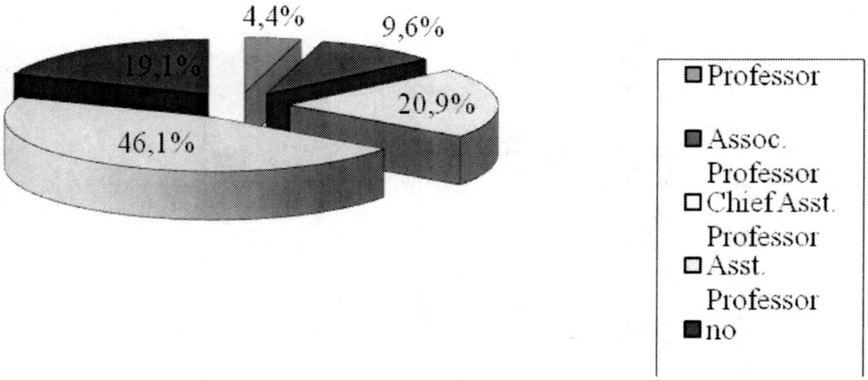

Figure 9.3 Distribution according to academic rank

for practitioners in university environment, respectively. Again, in the two groups, the third largest are the subgroups, in which neroticism is in 'average' degree.

Result comparison in the clinical-diagnostic categories 'practically healthy, 'preneurotic state' and 'marked neuroticism' according to the factors 'years of practical activity' and 'sex' reveals a mixed clinical picture. The group 'work up to 10 years' constitutes 37,69% of all participants. In them the number of practically healthy men (45,71%) and women (11,44%) predominates. In the group with length of service between 11 and 20 years, the number of the participants falling in the subcategory 'preneurotic state' (men – 18,75%; women – 31,25%) increases. In those whose practice is between 21 and 30 years, a decrease in the preneurotic states and dispersion of the results in 'mild', 'average' and 'severe' degrees in the category 'marked neuroticism' is recorded. The practically healthy are 40%, and 25% are in preneurotic state. In men and women there is a match in the values' mild degree of neuroticism' (5%). In 'average' degree, it is recorded in 15% of men and 5% of women. The 'severe' degree is recorded only in men (5%). I.e. the anxiety state, represented in the survey, excludes the development of complete burnout syndrome among lecturers and it rather determines a clinical picture of 'emotional satiation' and 'professional burnout'.

The results lead to the conclusion that the combined working conditions and the integration of influence of irritant stimuli for a prolonged period of time affects the individual health condition. The tendency for anxiety increase in the labour process inevitably affects the evaluation for job satisfaction, environment and conditions of work.

The highest satisfaction is recorded in lecturers working in a 'classical' educational environment of the medical university (53,12%). Confirmation of the conclusion that in the working environment, defined as 'traditional' for this type of profession, brings greater satisfaction than work in one, in which elements of two or more working environments are combined, is found in physicians occupied only with clinical practice. In them satisfaction is 52,39%. Although satisfaction with physicians, lecturers in university hospital is high (46,67%), it does not exceed 50% which is the indicator for satisfaction. In this group the highest percent of dissatisfied with labour conditions (30,00%) is recorded. The result comparison among the three studied groups reveals a statistically significant difference (χ^2= 24,566, df = 6, p = 0,000). It could be argued that evaluation of job satisfaction among physicians, lecturers in medicine is influenced by the factor 'training of students'. If university lecturers' professional realization is achieved through training of students who are the subject of their work, then for physicians, combining medical and educational activities, student training is a reason for concern and anxiety. I.e. teaching and participation in other type 'subjec-subject' relationships, which are beyond the standard for the clinical environment relationships between doctor and patient, is a serious psychosocial stressor for provoking anxiety.

The data analysis of the criterion 'not satisfied' also registers a difference (χ^2= 16,34, df = 2, p <0,000), which leads to the conclusion that satisfaction from teaching is a value dependent on the factor effectiveness of training in the context of the characteristics of the educational environment. Job satisfaction is a sum of the satisfaction from the efforts invested and performance in the medical, as well as in the educational activity. Lack of satisfaction caused by work in the educational environment influences the general evaluation for satisfaction from work in the complex working environment of medicine.

The presence of satisfaction in 'average' (36,52%) and 'low' (11,30%) degree, as well as of the inverse proportional correlation between satisfaction and burnout reveals, that it is more likely for the employees in a combined working conditions to speak about unlocking a process which is defined as 'emotional burnout'(Figure 9.4).

Determination of the psychosocial stressor 'student training' is confirmed by the answers of the survey conducted, where according to 53,33% from physicians- lecturers, working in the structures of the university hospital- creation of lecture courses, variants of semester and examination tests, checking students' presentations, papers etc., are activities done outside the clinic- in the

Figure 9.4 Jobs satisfaction

free time. The comparison of the answers of both groups lecturers of medicine confirms the statement that medical student training as a function of the intergrated type of work, is a factor for increased anxiety ($\chi^2 = 60{,}027$, df = 4, p<0,000). This places the integrated labour in the role of a psychoemotional stressor, originating from the nature of the performed practical activity.

Other specific micro-stressors, operating in the educational environment of the medical university and affecting anxiety and the feeling of satisfaction, are maintaining knowledge of a foreign language or learning a new one (71,70 % / n = 115), the difference in the theoretical knowledge in Bulgarian and foreign students, due to curriculum differences in secondary schools (85,40 %), irregular class attendance by the students (61,00%), difficulties with the language, experienced by foreign students in their educational process (56,10 %), change in the individual manner of expression, due to the language competence of the auditorium (53,70 %), consideration of the differences in cultures, represented at the university (34,10 %) etc.

Result analysis highlights an important characteristic of the educational environment in the medical university, which affects anxiety- the inetrculturality of the educational environment in medicine. This peculiarity transforms into a significant factor for inducing psychic tension, uneasiness and anxiety ($\chi^2 = 58{,}143$, df = 4, p<0,000). Interculturality is a predicator for more intense emotionality in the interactions carried out in the educational environment.

The focus on interculturality as a peculiarity in the medical training, transferred from the 'purely university' into hospital environment, places it in the position of a specific psychosocial stressor.

The answers of two statements- 'work exhausts me' (Figure 9.5(a)) and 'I hardly wait until the end of the university year' (Figure 9.5(b))- presuppose an increased risk from emotional exhaustion of the lecturer

A way for limiting the influence of micro stressors in the educational environment is rigidity. Rigidity is an inability for change in the mental attitude of the individual and lack of variability in behavior which is a prerequisite for

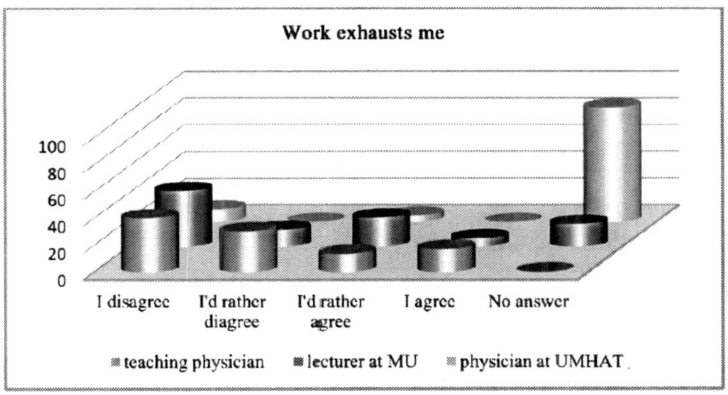

(a)

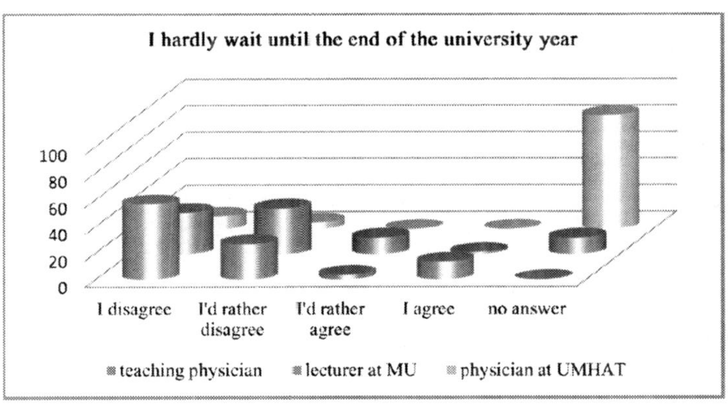

(b)

Figure 9.5 Emotional exhaustion in educational working environment (a) Work exhausts me (b) I hardly wait until the end of the university year.

lack of adaptability. (Stoytcheva, 2011). It can be assumed that in physicians with long-term teaching experience, rigidity is more likely to be unwillingness for change in the already established internal attitude to work with students. Anxiety with them is due to changes imposed by the conditions of the external environment- for example, reading lectures in a foreign language, changes in the curriculum because of new educational standards, requirements etc., then in younger lecturers apart from them, anxiety is provoked by work in auditorium, time organization, acquiring new competences etc. The statement that student teaching is a practical activity, emotionally loading their colleagues and and increasing their working activities, is shared by 90,47% of physicians who are not lecturers (n = 21).

Determining some of the basic factors for influence on anxiety among lecturers of medicine are grounds to be represented through a pattern of the specific psychosocial stressors provoking anxiety and leading to emotional overload and emotional exhaustion (Figure 9.6).

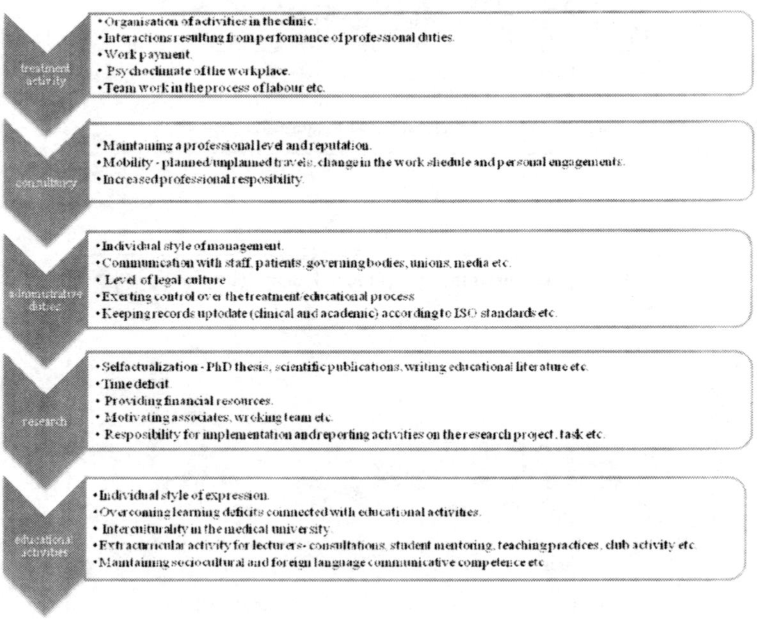

Figure 9.6 Exemplary model of the specific psychosocial stressors acting in the combined educational environment in medicine.

The attention paid to the three research tasks – diagnostics of the labour environment, determining specific psychoemotional stressors in it and tendencies for overcoming or increasing anxiety among employees, points to some conclusions.

- The anxiety state at the university hospitals reveals that the working conditions in them predispose to interactions which are emotionally intense. In this sense, the working environment contains prerequisites such as influence on the psychoclimate as well as on the anxiety level.
- Integration of elements from several working environments, combining activities, assumption of responsibility, excessive labour activity etc., generates an anxiety state and is a predictor for emotional exhaustion. These elements appear to be complex psychosocial stressors which contain specific tools and under their influence anxiety occurs.
- Anxiety and depressive states among lecturers at the university hospitals represent a picture of an initial process defined as 'professional flame out'. According to the study, the anxiety level excludes the complete forms of burnout syndrome. This is evidenced by the presence of satisfaction in 'average' and 'low' degrees among employees in combined working conditions.

References

[1] Asenova, R. Et al.. (2003). Psychoemotional 'exhaustion' of GPs in: First national conference for study and combat with stress, Plovdiv, p.207–212.

[2] Borisov, V., Ts. Vodenitcharov (2000). Realities in the national reform, Sofia, Georgieva, M.T. Vladeva (2012). Evolutionary basic for perception of individual feelings, in: Psychology journal, Nov. 2012, available at – http://psychologiournalbg.com.

[3] Zhivkova, H. (1996). Empathy in the medical profession in: Bulgarian Journal of Psychology, issue 1, p. 35–56.

[4] Ivanova, A. (1992). Content of work and value orientation in the labour activity, in: Bulgarian Journal of Psychology, Nō 4, p. 41–56.

[5] Ivanova, N. Et al. (2003). Professional- work stress In GPs and physicians, in: First national conference for study and combat with stress, Plovdiv, p. 241–250.

[6] Ilieva, S., V. Naydenova (2007). Stress and healthcare attitude in the university environment, in : Sofia University 'St. Kliment Ohridski' yearbook, Psychology book, t. 99, p. 63–88.

[7] Nunev, S. (1999). Professional adaptation of the young social pedagogue and the syndrome of 'emotional flame out' ", in: Education, book. 6, p. 43–49.
[8] Selie, H. (1982). Stress without distress, Science and Art, Sofia, p. 22–23.
[9] Stoytcheva, Katya, Dimitar Shtetinski, Kalina Popova (2011). Study of the personality rigidity with the survey of Ivan Paspalanov (π.P. – 1), in: Psychology magazine, issue. 3–4, p. 136–142.
[10] Tilov B., B. Tornyova, D. Stoyanov (2012). Psychological climate as a risk factor for burn out incidence, in: Personality, psychoclimate and burn out professional syndrome. Guidelines on diagnostics and prevention of burn out syndrome in healthcare professionals, p. 61–78.
[11] Tulevski, B. (1986). Social factors and stress in the medical practice: Conceptual model. Problems of social medicine. National institute in social medicine, Sofia, , p. 34–42.
[12] Hanin, Yu. (1978). Anxiety study in sports in: Psychology questions, No 4, p.94.
[13] Hristov, Zh., L. Tomev, D. Kircheva, N. Daskalova, T. Mihaylova, Z. Naydenova (2003). Psychoemotional and health consequences from stress among employees in education, healcare and state administration-pilot project. National conference 'Stress', Sofia, p. 282–289.
[14] Hristov, Zh. (2004). Stress, stressors and stressogenic situation, in: Physical medicine, rehabilitation and health, issue 2, p. 11–16.
[15] Hristov, Zh., D. Stoyanov (2009). Management psycholofy in healthcare. Textbook for MAs in health management, Kameya Design Ltd, Plovdiv.
[16] Hristov, Zh. (2009) Stress in workers and staff in the actual economic sector in Bulgaria, PhD Thesis
[17] Hristov, Zh.., E. Gospodinova, D.Hristova, T. Taneva (2010). School management, anxiety and health of the teacher. Public health and health care in Greece and Bulgaria, ?. 212–218.
[18] Tsenova, B. (1993). Burn out phenomenon, in: Bulgarian Journal of Psychology, issue 4, p. 35–56.
[19] Tsenova, B. (1996). Professional stress and mental health of teachers. – ч.I, Bulgarian Journal of Psychology, No 4, p. 53–55.
[20] Tsenova, B. (1998). Psychological terror on the work place and organization, in: Challenges in human relationships, p. 54–64.
[21] Tsenova, B., E. Ivanovich (2002). Stress at work. Tsenova, B., E. Ivanovich (2002). Stress at work. What is required by the employer,

in: Safety and health in work, issue 6, p. 64—77. What is required by the employer, in: Safety and health in work, issue 6, p. 64–77.

[22] Flin, R.&C. Burns et al. (2006). Measuring safety climate in health care Qual Saf Health Care, p. 109–115.

[23] Bem, D.J. (1972). Self-perception theory. In L. Berkowitz (Ed.), Advances in experimentalsocial psychology (Vol. 6, pp. 1–62). New York: Academic Press.

[24] Burnout : Illness or symptom? Kapfhammer HP. Klinik für Psychiatrie, Medizinische Universität Graz, Auenbruggerplatz 31, 8036, Graz, Österreich.

[25] Laird, J. D. (1974). Self-attribution of emotion: The effects of expressive behavior on the quality of emotional experience. Journal of Personality and Social Psychology, 33, 475–486.

[26] Laird, J. D. (1984). The real role of racial response in experience of emotion: A reply to Tourangeau and Ellsworth, and others. Joumal of Personality and Social Psychology,47, 909–917.

[27] James, W. (1890). Principles of psychology. New York: Holt.

10

The Impact of Burn Out on Quality of Life in the Context of Professional Realization in Medicine

Slavka Toshkova-Hristozova PhD,
Department of Language and Specialized Training, MUP

Drozdstoy Stoyanov MD, PhD, Full Professor of Psychiatry,
Medical Psychology and Person Centered Medicine,
Faculty of Medicine, MUP, Vice Chair Executive Committee PSIG,
Royal College of Psychiatrists, Visiting Fellow, University of Pittsburgh

Kaloyan Haralampiev PhD, Associate Professor,
Department of Sociology, Faculty of Philosophy,
University of Sofia "St. Clement of Ohrid"

In the recent years in scientific literature a considerable attention is paid to the socially significant problem of the mental 'satiation' and the nervous exhaustion among the people practicing professions subordinate to the relations of the type 'human- human'. The data published correlate directly with the problem of emotional burnout. In this context, each new study dedicated to the psychological overload connected with the execution of professional duties, professional development and career growth, is an aspect in the clarification of the nature of the burn-out syndrome as well as of defining new basic psychosocial stressors predisposing its manifestation in the specific professional environment.

The strife for professionalism is an affirmation in the profession through improving knowledge, acquiring competences etc. and they accompany the professional pathway of specialists from different fields. But for those who practice an occupation with 'subject-subject' nature, the quality of the performed professional activities is influenced by the impact the nature of the practical activities has on the individual psyche.

Drozdstoy St. Stoyanov (Ed.), New Model of Burn Out Syndrome: Towards Early Diagnosis and Prevention, 131–146.

Professional development and professional growth are defined as a process of professionalization. In its essence professionalization is a complex entity which has time parameters. It begins with the mastering of the profession and ends with the termination of its practice. Professional build up goes through different stages, it is connected with experiencing difficulties, over expenditure of psychic energy, intensification of emotional reactions etc. The prolonged emotional stress in the 'subject-subject' professions combined with the strife to maintain high professionalism at work presupposes reaching a point where a reduction in the emotional capacity of a person begins, thereafter a demotion of the capacity for work and reduction of work efficiency begin. Loss of initiative for self actualization, of the strife for improvement in the personal professional achievements is reported (Stoyanov, 2012). What unites the above mentioned professions is a high level of psychoemotional stress, predisposing emotional overload and a disruption in the individual adaptation mechanism.

The decrease of the individual adaptation ability limits the options for management with the negative pressure exerted on the side of the psychoemotional stressors in the immediate professional environment. Not only the efficiency and effectiveness of work are affected but also the work post, the achieved position, the acquired social status, the material prosperity etc. are endangered. Other negative consequences are also chronic tiredness, low self-esteem, declining health, difficulties in coping with every-day problems. The declining health status, a consequence of the professional activity, influences job satisfaction and thence the evaluation of quality of life. The transfer of anxiety and problems from work to the home, their experience in family and friendly environment, the emotional intensity of interpersonal relationships, caused by the declining psycho climate in the professional surroundings etc. have 'an impression' on the individual behavior and reflect on the manner of life and its quality.

The term 'manner of life' derives from the social sciences and characterizes the conditions and peculiarities inherent to everyday life. It encompasses all existing spheres of social life- from social organization and economic problems of the socium to the individual living conditions, free time, satisfaction etc. (Encyclopedia, 1974). In sociology the term 'lifestyle' is used. 'Standard of life' is also encountered containing the understanding for the individual choice of the personality due to which the understanding for many styles of life is defended (Toffler, 1970).

When the balance between the requirements for professionalism and psychic resources of the individual is disrupted, the undertaking of many working activities puts to the test the psychic stability of the personality.

The excessive work load, psychic and physical overload, tension, emotional exhaustion are the causes for loss of motivation, decrease of interest in the practiced activities, the occurrence of lack of initiative (Stoyanov, 2012). I.e. the manner of life, its dynamics and style are defined by the personality and reflect its individual understanding for quality.

One of the definitions for quality of life is its representation as 'subjectively derived characteristic of prosperity, result of the summary evaluation of individual and clinically significant domains' (Stoyanov, 2012). Quality of life is influenced by the discrepancy between expectations and the actual achievements. Its characteristic feature is the feeling for prosperity in the light of individual satisfaction of needs. Physical and mental health, social relationships, economic factors pertain to the quality of life. In this sense the influence of macro and micro-stressors from the external and internal environment combined with prolonged psychic overload are prerequisites for negative influence on health of the employee and his/her evaluation of quality of life.

The general feeling of healthiness, worries and anxieties as well as emotional well-being are essential in defining the quality of life in its entirety (Balabanov, 2006). This is the reason why in the recent years the problem occupies a place among the contemporary issues connected with public health and healthcare organization. Quality of life becomes a criterion of evaluation of the healthcare services provided, of the character and the extent to which the intervention performed and the treatment carried out have an effect on the disease (Bullinger, 1997). Quality of life depends on the factors daily activity, role of the family, physical abilities, self-service, entertainment, socialization etc. (Grigorov, 2003). In medico-social aspect it presupposes application of biopsychological approach to the patient's health. The approach is aimed at the early detection and influence on the psychological problems and social restrictions in the context of the disease. It is connected with health as a universal value and as a indicator for quality of life (Health – related quality of life – HRQOL).

Quality of life expresses the subjective evaluation of life and is comprised of the combination of physical, mental and social health. It consists of three hierarchical levels- Intrapersonal (physical and mental health, independence and concept of oneself), interpersonal (family life, friendships, social relationships and support, religion, spirituality) and extra personal (spare time, education, environment outside home etc.) (Viteva, 2006). Quality of life is provoked by the negative influence of psychosocial factors which in the professional environment play the role of micro-stressors leading to weaker professional realization, lack of satisfaction from the social life, reduced self-esteem, worsened psycho climate, increased anxiety, manifestation of depressions etc. (Tsenova, 1996).

The influence on personality on the side of micro-stressors is much stronger when the environment in which it achieves professional realization, does not fulfill the criteria for 'classical' environment for practicing a specific activity. An environment which is associated with the training of specialists, whose education requires mastering practical activities in a clinic, laboratory, industrial unit etc., is a combined educational environment. With a view of the similarity between educational and working activities, the lecturer and students, participating in a common educational process, are participants in a common work process. It takes place at the university's hospital clinics, which in the context of the educational environment in medicine, has the characteristics of a combined working environment -hospital and teaching.

The professional realization of the lecturer in medicine is focused on the professional development in the field of medicine as well as on mastering competencies obligatory for practicing teaching activities in the field of medical education. In practice, labour activity, self actualization, additional work-load incl. working overtime etc. limit the free time. The overload at work combining medical and teaching activities are a prerequisite for emotional 'satiation'. It imposes working at an accelerated rate and provokes psychic 'compression' on the personality and thus limits the time for rest and recuperation. The pressure which the total complex of activities exerts, presupposes investing more emotional and physical effort, greater work load, which cause psychic exhaustion and lead to decline in the quality of life.

Among the most important factors determining quality of life are emotional well-being, general feeling of health (hypertension and diabetes), worries and anxiety, social and professional well-being, physical and mental health (Balabanov, 2006). Their universal validity transforms them in basic criteria for the study of quality of life incl. that of the lecturer in the educational environment of medicine. In 2012 a study was conducted encompassing representatives of all medical universities in Bulgaria- Plovdiv, Sofia, Varna and Pleven (n= 115). Its goal was to elaborate on a general idea of the quality of life in the context of the professionalization of the physician as it reports on the presence of emotional exhaustion and the first manifestations of the burn-out syndrome. Separate study tasks are evaluation of satisfaction, anxiety and the initial display of burn-out syndrome among lecturers at Medical University – Plovdiv. The method of study include a survey, test on satisfaction from the working conditions and the perspective for professional development, test for anxiety rating (of T.Tashev) and test for burn-out presence (K. Maslach).

Quite often in his professional activities the lecturer of medicine finds himself in stressful situations. They arise from the nature of the work in the clinic and the relation physician- patient as well as from the nature of the teaching of medicine and the relation lecturer - student. As a result from the complex interaction a prerequisite for increased emotionality in the relationships, emotional pressure on the personality, psychic tension and frustration arises. The intensity and duration of the negative effect of the stressors lead to the development of metabolic disorders; they affect the cardio-vascular system, endocrine system and lead to carcinoma development etc. Apart from the influence on physiologic and mental levels, development of negative characteristics of behaviour are reported- aggression, capricious behaviour, depressive state, alcoholism, drug abuse etc. (Nunev, 1999). Therefore one of the study tasks of the survey is to record the presence of hypertension and diabetes as the most common metabolic disorders. With the reservation that their occurrence is not bound only with the effect of stressors in the professional environment, these diseases influence the practical activities and quality of life.

The survey results show that 40% of physicians working in a combined environment of the university hospitals (UMHAT and MU) suffer from hypertension. The comparison of the result with that of their colleagues from the medical university (MU) reveals that among those working only in the academic environment 32,80% suffer from hypertension. In the other control group-physicians at UMHAT and general practitioners (GP), hypertension is recorded in 9,50%. The complex nature of the labour activity deduces the nature of the combined work in a factor for negative influence on the health condition of the lecturer. The statement is proven by the non-parametric test of Person ($\chi^2 = 53{,}256$, df = 4, $P<0{,}000$).

12,20% from the participants in the survey indicate that their quality of life is influenced by the occurrence of diabetes. The comparison of the groups according to years of teaching practice reveals that diabetics are: 9,10% from those who have worked up to 5years, 35,30% up to 15 years, 7,70% up to 25 years and 9,70% more than 25 years. Among the physicians who work only in the hospital environment of UMHAT and the healthcare network (GP), the disease is recorded in 4,80%. The results on years of teaching practice indicate the duration of the professional realization in the combined working environment in medicine as a factor for triggering a metabolic disease ($\chi^2 = 64{,}074$, df = 8, $P<0{,}000$).

According to the opinion expressed in the survey 41,70% of teachers do not accept that the level of the work tasks performed is influenced by their health condition. In a low degree hypertension and diabetes affect the work of 26,10% of surveyed and in a considerable degree 7,80%. These groups include mainly employees with teaching practice around and more than 25 years (Table 10.1). The comparison of groups determines another statistical difference (χ^2 = 68,419, df = 12, P<0,000) according to which the level of professionalism including the quality of medical education is influenced by the factor individual health expressed by working capacity, psychic state and energy put in the work of the lecturer.

The comparison of the results represents another important factor for metabolic disease incidence and development (χ^2 = 96,800, df = 12, P<0,000) – the peculiarities of the workplace as a factor for health and level of practical activities. The result is influenced by the peculiarities of the internal environment characteristic for every medical university as well as by psychosocial stressors of the external environment- town, geographic features, financial security of activities etc. affecting the standard of life in the different settlements. The conclusion stated is that with lecturers working in a combined working environment in medicine in Plovdiv the metabolic disease affects in 'low' degree 47,61% of surveyed and in 'considerable' 9,53%. Therefore more than half of the lecturers in UMHAT are affected. This presupposed the development of two tendencies- in degradation of the quality of the work performed or in preserving the professional level at the expense of individual health, exhaustion of the psychic resource and decline of the quality of life.

The deduced dependencies between health, professional environment and quality of life of doctors teaching medicine and practicing in a combined working environment are a further confirmation of the fact that the working conditions, the nature of the profession, psychoemotional stress connected with its practice are basic constructions in the 'building' up of an evaluation of the quality of life and the individual understanding for well-being and satisfaction.

One element in the evaluation of quality of life is the dependency work load- time for rest and recovery. The excessive labour activity and sleep deprivation, entertainment and relaxation, impairs quality of life expressed in limitation of the social contacts, redistribution of family duties, shortening of the time for communication with the children, parents, friends; focusing on work problems, self-isolation, falling into a depressive state etc.

Table 10.1 Influence of Disease on Work - Distribution in Years of Practice (n = 115)

Degree of influence	Lecturers at UMHAT and MU												UMHAT and GP			Total		
	Up to 5 years			Up to 15 years			Up to 25 years.			More than 25 years.								
	Num.	P [%]	Sp.	Num.	P [%]	Sp.	Num.	P [%]	Sp.	Num.	P [%]	Sp.	Num.	P [%]	Sp.	Num.	P [%]	Sp.
No influence	19	57,60	11,34	11	64,70	14,41	9	69,23	15,38	9	29	15,13	–	–	–	48	41,74	7,12
In a low degree	11	33,30	14,21	3	17,60	21,99	4	30,77	23,08	11	35,5	14,43	1	4,80	21,38	30	26,09	8,02
In considerable degree	1	3,00	17,06	2	11,80	22,81	–	–	–	4	12,90	16,76	2	9,50	20,73	9	7,83	8,95
No answer	2	6,10	16,92	1	5,90	23,56	–	–	–	7	22,6	15,81	18	85,70	8,25	28	24,34	8,11
Total	33	100		17	100		13	100		31	100		21	100		115	100	

In the comparison of the three generalized groups- lecturers working in a combined working environment, working only in MU and physicians practicing in UMHAT and GPs it is reported that 36,70% from those who practice an activity in UMHAT and MU do not have enough free time for rest. 23,30% from the lecturers note that they have an extra job- i.e. influence of the economically defined external macro-stressor is registered (Figure 10.1)

The reported state among lecturers at MU and UMHAT is a cause and prerequisite for the study of the problem- are there initial manifestations of a burn-out symptom among the three studied groups. The diagnostics of burn-out syndrome is based on the qustionnaire of Christina Maslach MBI containing 22 questions distributed in three subscales – emotional exhaustion (EE)- 9 questions, depersonalisation (DP) – 4 questions, and personal accomplishment (PA) – 8 questions. According to the methodics of reporting, the results obtained in the separate subscales are the following:

Highest degree of emotional exhaustion is reported in the group which practices only therapeutic activities (76,19%). The lowest is in the lecturers who work in the theoretical departments of the medical universities (56,25%). Probably the 'average' degree of emotional exhaustion among the employees in the combined medical environment is influenced by the training of students and work with young healthy people. It is assumed that work with them compensates emotional exhaustion connected with the nature of the therapeutic activities – work with patients in a hospital environment (Table 10.2). The comparison of the results of the different groups reveals that there is a correlation- i.e. the subject of the labour activity is a factor in emotional exhaustion ($\chi^2 = 32{,}984$, df = 4, $P<0{,}000$).

The second category- depersonalization (dehumanization) reports a negative perception and negative attitude to oneself and the others. Negativism in the evaluation of others and oneself affects satisfaction- i.e. it influences directly the evaluation of quality of life. (Figure 10.2)

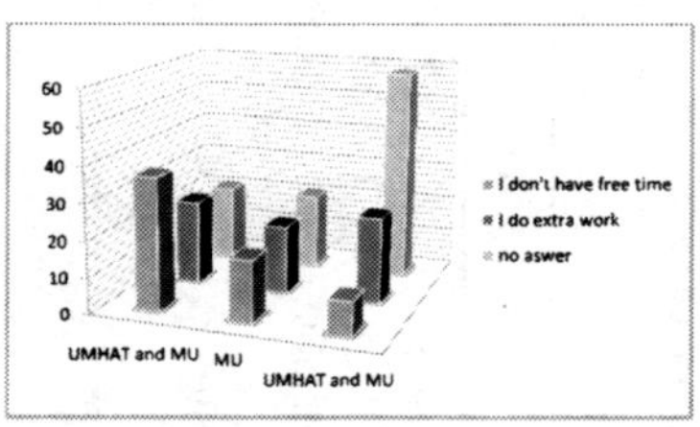

Figure 10.1 Predictors for Emotional Exhaustion

Table 10.2 Degree of Emotional Exhaustion According to the Scope of Activity (n = 115)

Degree of exhaustion	Medical and teaching			Teaching			Treatment			Total		
	Num.	P [%]	Sp.	Num.	P [%]	Sp.	Num.	P [%]	Sp.	Num.	P [%]	Sp.
Low	5	16.67	16,66	36	56.25	8,27	3	14.29	20,00	44	38.26	7,30
Average	13	43.33	13,74	16	25.00	10,80	2	9.52	21,00	31	26.96	8,00
High	12	40.00	14,14	12	18.75	11,30	16	76.19	11,00	40	34.78	7,50
Total	30	100		64	100		21	100		115	100	

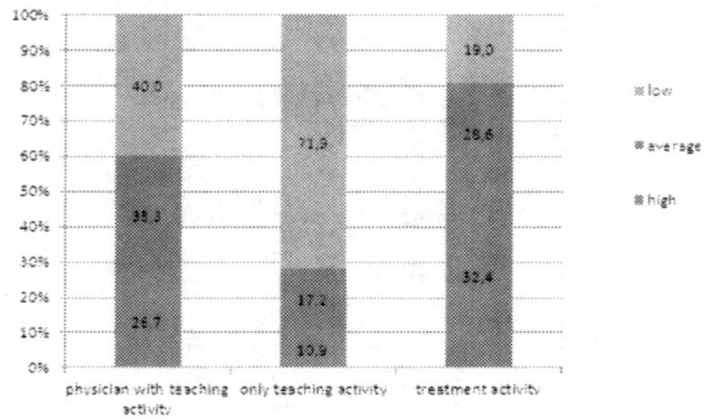

Figure 10.2 Degree of Depersonalization According to the Kind of Activity

The summarized results reveal that depersonalization in degree 'low' is 53,91%, and that in the degrees 'average' and 'high' are respectively 23,48% for the first and 22.61% for the second. The essence of labour – whether bound by implementation of activities relating only to one profession or combining activities inherent to more than one profession is a factor for depersonalization (χ^2 = 24,501, df = 4, P<0,000).

The comparison of results according to workplace and reporting the peculiarities on the level of uniersity, of town, of geographical region et., shows overlapping in the 'average' degree as with emotional exhaustion as well as in depersonalization of the employees in the combined working environment of UMHAT at MU- Plovdiv (47,60%). In both subscales the lecturers working only in the educational environment of the medical university form groups with degree' low'. The relative share for MU-Plovdiv is 56,00%, and for the groups of lecturers from Sofia, Varna and Pleven it is 86,49%. In the group of lecturers working at UMHAT and GPs, depersonalization is reported in 61,11% – i.e. they again indicate values pertaining to 'high' degree. The fact that the peculiarities of the workplace in the context of town, standard of life etc. are factors for depersonalization of the personality is proven by the correlation χ^2 = 45,096, df = 8, P<0,000).

The third subscale of the MBI test presents personal accomplishments as a criterion for defining the presence of burn-out syndrome among the academic medical lecturers (Table 10.3). The reduction in the personal accomplishments expresses the reduction in the interest for professional development and refers to working capacity. The highest working capacity is reported to the employees only in the educational environment of MU (62,50%). The academic medical lecturers practicing in a combined working environment form the group with

Table 10.3 Personal Accomplishments- Distribution According to Kind of Activity (n = 115)

Personal accomplishments	Medical and teaching			Teaching			Treatment			Total		
	Num.	P [%]	Sp.	Num.	P [%]	Sp.	Num.	P [%]	Sp.	Num	P [%]	Sp.
High	12	40,00	14,14	40	62,50	7,65	5	23,80	19,00	57	49,57	6,60
Average	7	23,33	15,98	16	25,00	10,8	8	38,10	17,00	31	26,96	8,00
Low	11	36,67	14,52	8	12,50	11,7	8	38,10	17,00	27	23,47	8,20
Total	30	100		64	100		21	100		115	100	

'average' degree of working capacity (40.00%). Among the employees only of UMHAT and general practitioners there is concurrence in the degrees 'average' and 'low'(39,09%). The dependency between the two types of activity and professional development is reported. The education of students and work in an academic environment are a factor for fostering the interest in professional development and motivator for development in it (χ^2 = 14,147, df = 4, P<0,007).

The conclusion is confirmed by the results obtained in the distribution by age teaching practice (Table 10.4). It appears that the working capacity among lecturers is sustained 'high'in the whole period of teaching practice. In comparison to them, the strife for professional development of physicians, who are not lecturers, is in 'low' degree (42,90%). The comparison among the groups of physicians represents student training as a motivator for professional development and factor simulating professional recognition in the field of medical education (χ^2 = 19,119, df = 6, P<0,014).

The presence of 'average' degree of emotional exhaustion among lecturers, working in the clinical base of the medical university, as well as the 'high' degree of personal accomplishments in lecturers shows that we cannot speak of professional burnout among the lecturers in Bulgarian medical universities. Initial process of dehumanization is recorded.

The survey results give us the grounds to state that in Bulgarian medical universities and in their the clinical base registers a picture which is close to the initial forms of burnout, more commonly referred to as 'burning'. The presence of emotional 'flame out' is a prerequisite for the next step- emotional 'burnout', originating from the nature of the labour activity. This imposes the prevention of burnout syndrome in the risk groups medical professionals, working at UMHAT and the healthcare network. The establishment of a prevention programme presumes reporting the current condition.

Analysis of the presented data in the context of quality of life of the academic medical lecturer reveals that it can be can be claimed that in Bulgarian medical universities and the clinical base in them a situation is reported which is similar to the initial phases of burn-out known as 'burning'. The presence of emotional flame out is a prerequisite for the next step-emotional burn out caused by the nature of the labour activity. This imposes prevention of burn out syndrome in the risk groups of medical specialists working in UMHAT and the healthcare network. The establishment of a programme for prevention presupposes accounting of the present state.

At the time of conducting the survey the recovery of mental overload and emotional exhaustion is in the active form- mostly travelling. Going abroad,

Table 10.4 Personal Accomplishments- Distribution in Years of Length of Service (n – 115)

Degree	Lecturers												Physicians UMHAT and GPs			Total		
	Up to 5 years			Up to 15 years			Up to 25 years.			More than 25 years.								
	Num.	P [%]	Sp.	Num.	P[%]	Sp	Num.	P[%]	Sp	Num.	P[%]	Sp	Num.	P[%]	Sp	Num.	P[%]	Sp
high	17	51,50	12,12	7	41,20	18,6	11	84,60	10,88	18	58,10	11,60	4	19,00	16,76	57	49,60	6,62
average	7	21,20	15,45	5	29,40	20,37	1	7,70	26,66	10	32,30	14,80	8	38,10	15,40	31	27,00	7,97
low	9	27,30	14,85	5	29,40	20,37	1	7,70	26,66	3	9,70	17,10	9	42,90	15,13	27	23,50	8,16
total	33	100		17	100		13	100		31	100		21	100		115	100	

trips around the country, visits outside the settlement are a preferable way for relaxation and 'emotional charging'. This is shared among 16,70% of the lecturers from a combined medical working environment, among 18,75% of those wortking only in MU and among 23,81% of those practicing therapeutic activities. In those practicing combined labout there is and overlapping of the categories ' travelling' and 'interest club'(16,70%) (Figure 10.3).

In the three studied groups the highest relative shares are in the category 'no answer' (Table 10.3). The highest value is for the group practicing only medicine (61,91%). Here again we register the indirect influence of the external psychosocial factors such as remuneration of labour, organisation of activities different for the doctors working in a combined working environment at UMHAT and in GPs. The comparison of the results in the three groups deduces a statistical difference (χ^2 = 32,681, df = 16, P<0,008), which presents the organization of activities connected with professional realization as a factor for influence on the quality of life.

The results from the survey justify the following conclusions:

1. The health of the academic teacher at UMHAT and MU is affected by the nature of labour expended in the combined working environment and possessing a complex nature.
2. The duration of work in the combined working environment of UMHAT for the academic lecturer is a prerequisite for triggering metabolic disorder.

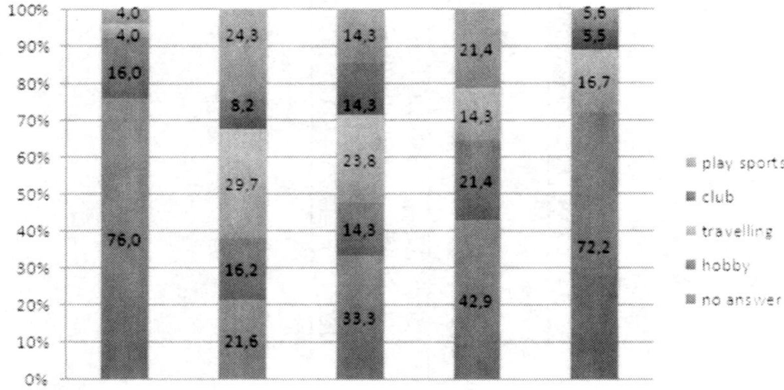

Figure 10.3 Management of Psychoemotional Stress

3. The conditions and peculiarities of the work place turn it into a factor for influence on health.
4. The level of professionalism incl. the quality of the educational process in medicine is affected by the factors individual health, working capacity and energy source of the lecturer.
5. Organisation of labour activity is a factor for defining the evaluation of quality of life.

The conclusions made by the study are in the context of professional practice and self – actualization of the academic lecturer of medicine. They give us the grounds to claim that practicing professions with a complex nature especially ones combining elements of 'subject-subject' professions are a factor in emotional exhaustion and emotional burnout of personality.

References

[1] Balabanov,P., Z. Zahariev. The quality of life of Bulgarian patients with focal and generalized epilepsy, Bulgarian Psychology Journal, issued by Bulgarian Nerological Society, vol. VI, copy 1, 2006, p. 38–44.

[2] Great Soviet Encyclopedia, vol. 18, Moscow, 1974, p. 217–218.

[3] Viteva, E., Z. Zahariev. The quality of life in patients with epilepsy. Bulgarian Neurology Journal, vol. 6, copy 3, 2006, p. 111–115.

[4] Grigorov, F. Quality of life. Evaluation of the quality of life in patients with cardio-vascular diseases, Bulgarian Cardiology Journal, year book IV, copy 4, 2003, p. 72–73.

[5] Nunev, S. Professional adaptation of the young social pedagogue and the syndrome of 'emotional burnout', in: Education, book 6, 1999, p. 32–42.

[6] Stoyanov, D., M.Stoykova. Burnout syndrome: origin and significance in medicine, in: Personality, psychoclimate and syndrome of the professional burnout, „East-West, Sofia, 2012, p. 26.

[7] Tsenova, B. Professional stress and psychic health of teachers- part 1, in: Bulgarian Psychology Journal, book 4, 1996, p. 46–72.

[8] Bullinger, M.: Health related quality of life and subjective health. Overview of the status of research for new evaluation criteria in medicine. Psychosom. Med. Psychology, 1997, book 47, p. 76–91.

[9] Toffler, A. The Future shook, London, 1970, p. 306.

11

Study of Burn Out Among Employees in Penitentiary System

Stanislava Harizanova MD, PhD, Assistant Professor,
Department of Hygiene and Ecomedicine,
Faculy of Public Health, Medical University of Plovdiv

Nonka Mateva PhD, Associate Professor of Social Medicine,
Mathematician and Statistical Advisor,
Department of Health Management, Health Economics and General Practice,
Faculty of Public Health, Medical University of Plovdiv

Tanya Turnovska MD, PhD, DSc, Full Professor,
Department of Hygiene and Ecomedicine,
Faculty of Public Health, Medical University of Plovdiv

11.1 Introduction

Working in a prison as a correctional officer is one of the most stressful, challenging and demanding occupations (Lambert et al., 2007). Workers are placed in potentially dangerous circumstances and are frequently exposed to intense situations that can often lead to emotional exhaustion and interpersonal stress (Garner, 2008; Maslach et al., 2001; Toch, 2002). Armstrong and Griffin (2004) regard prisons as unique working environments as very few other institutions are charged with the primary duty of supervising and securing a population that can be unwilling and potentially violent. Research has found that the perceived dangerousness of the job, as a result of threats and inmate violence is a significant cause of stress for many correctional staff. One possible outcome of prolonged or chronic stressors at the work place is burn out. Higher levels of burn out are associated with shift work, a lack of stimulation on the job, low job autonomy, a lack of participation in decision making, low supervisor support, frequent contact with suspects, emotionally demanding and physically dangerous work situations (Lynch, 2007).

Drozdstoy St. Stoyanov (Ed.), New Model of Burn Out Syndrome: Towards Early Diagnosis and Prevention, 147–160.

Personality traits are also related to burn out (Storm & Rothmann, 2003a). Neuroticism, extraversion, external locus of control, job-distance inability, existential frustration and ability of love prove to be personality traits with an impact on the development or the presence of burn out (Bühler & Land, 1990).

Correctional officers are enmeshed in a unique work environment.

11.2 Aim

The main purpose of this study is to examine burn out syndrome among Bulgarian correctional officers. One of the focuses of this research is to examine whether personal and demographic characteristics are related to burn out.

11.3 Materials and Methods

The all available staff at prisons in the Regional Prison of Pazardzhik and the Regional Prison of Sliven was surveyed. The prison in Pazardzhik for men inmates encompasses the building of the prison itself and two open type correctional communities. In country-regionBulgaria there is only one prison for women and one correctional facility for minor girls, both located in Sliven. The primary information was collected among 201 people (100 were employees of the Regional Prison of Pazardzhik and 101 – of the Regional Prison of Sliven). The only qualification in the sample selection was that the employee has direct contact with inmates. The front page of the survey explained that participation was voluntary and anonymously, and the results would be kept in strict confidence.

Three major constructs – demographic variables, burn out assessment and personality characteristics – were included in the used questionnaire. Six demographic characteristics were selected – gender, age, marital status, education, tenure, and job position. The most widely used and validated instrument for the measure of burn out is the Maslach Burn Out Inventory (MBI) developed by Maslach and Jackson (1986). The Bulgarian version of the MBI performed by B. Tzenova was used to measure the three core dimensions of burn out – emotional exhaustion, depersonalization and reduced personal accomplishment. Eysenck Personality Questionnaire (1975) is a questionnaire to assess the personality traits of a person.

The statistical program SPSS version 17.0 was used for the statistical analysis. A descriptive analysis (mean, standard error of means) was conducted on the sample, followed by a Student t-test, one-way ANOVA and Pearson correlation coefficients. A p-value < 0.05 was considered statistically significant.

11.4 Results

Table 11.1 reports the major demographic characteristics of the survey respondents.

The means and standard error of means for three MBI subscales are presented in Table 11.2. In Figure 11.1 we present the levels of the components of burn out among correctional officers.

Table 11.1 Descriptive Statistics of Demographic Variables

Measure	NumberN	Percent %	Mean ± SE
Gender			
Female	87	43.28	
Male	114	56.72	
Age			
Continuous years			41.32 ± 0.54
Education			
College	124	61.69	
University	77	38.31	
Marital status			
Married	146	72.64	
Divorced	22	10.95	
Widowed	6	2.99	
Single	27	13.43	
Position			
Officer	141	70.15	
Inspector	60	29.85	
Tenure			
Years of service at the prison			11.71 ± 0.50

Table 11.2 Means and Standard Error of Means for MBI Subscales

MBI subscales	No of cases	No of items	Mean	SE
Emotional exhaustion	201	9	15.19	0.71
Depersonalization	201	5	7.98	0.38
Personal accomplishment	201	8	30.09	0.43

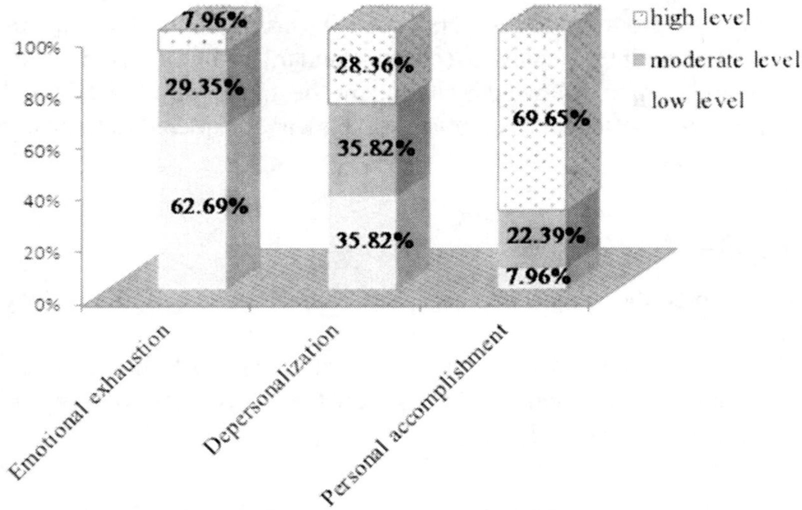

Figure 11.1 Levels of the components of burn out among respondents

We used a Student t-test and one-way ANOVA to check the significance of the mean differences of burn out subscales between the employees according to their demographic characteristics (Table 11.3). A t-test analysis showed that female staff generally report higher levels of emotional exhaustion and personal accomplishment than do male staff. The policemen have higher levels of depersonalization than policewomen.

As can be seen in Table 11.4 the introverted and emotionally unstable correctional officers indicate higher emotional exhaustion, depersonalization and reduced personal accomplishment than the extroverted and emotionally stable members of staff.

Based on the results reported in Table 11.4, we used a Pearson correlation between the burn out subscales and personality traits for measurement how changes in one variable is associated with changes in another. We found a strong significant correlation between all three components of burn out and personal characteristics (Table 11.5).

11.5 Discussion

Burn Out is a problem among correctional staff. Keinan and Maslach-Pines (2007) report that the correctional employees in their study had much higher levels of burn out than the levels found in the general population, even higher than police officers. In Bulgaria at present burn out among

Table 11.3 Impact of Demographic Characteristics on Three Burn Out Dimensions

MBI subscales Demographic characteristics	EE Mean ± SE	DP Mean ± SE	PA Mean ± SE
Gender			
Male	13.61 ± 0.94**	8.41 ± 0.59*	31.11 ± 0.58**
Female	17.26 ± 1.04**	7.65 ± 0.50*	28.76 ± 0.60**
Age			
< 30	10.75 ± 1.55	6.83 ± 1.43	31.42 ± 1.35
31 – 40	14.47 ± 1.18	7.90 ± 0.59	30.15 ± 0.70
41 – 50	16.38 ± 2.19	8.46 ± 0.62	29.43 ± 0.65
> 51	16.09 ± 2.19	7.22 ± 1.06	31.52 ± 1.18
Educational level			
University	15.61 ± 1.11	8.42 ± 0.65	30.13 ± 0.69
College	14.93 ± 0.92	7.71 ± 0.46	30.07 ± 0.54
Marital status			
Married	15.01 ± 0.85	8.12 ± 0.44	30.00 ± 0.51
Divorced	16.50 ± 1.99	7.64 ± 1.07	30.95 ± 1.49
Widowed	14.17 ± 4.21	4.67 ± 1.96	30.17 ± 1.96
Single	15.30 ± 1.90	8.26 ± 1.18	29.89 ± 0.93
Rank			
Officer	15.77 ± 0.88	8.33 ± 0.46	29.95 ± 0.53
Inspector	14.14 ± 1.38	7.23 ± 0.79	30.80 ± 0.79
Years of service at the prison			
< 5	15.17 ± 1.56	7.94 ± 0.87	30.42 ± 0.83
> 5	15.20 ± 0.79	7.99 ± 0.42	29.99 ± 0.50

EE = emotional exhaustion, DP = depersonalization, PA = personal accomplishment

* p-value is < 0.05.

** p-value is < 0.01.

Table 11.4 Impact of Personality Characteristics on Three Burn Out Dimensions

MBI subscales Personality trait	EE Mean ± SE	DP Mean ± SE	PA Mean ± SE
Extraversion			
Low	23.24 ± 2.89***	11.76 ± 1.35***	27.29 ± 1.57***
High	10.82 ± 0.97***	6.35 ± 0.54***	32.15 ± 0.70***
Neuroticism			
Low	14.16 ± 0.77***	7.51 ± 0.42***	30.52 ± 0.48***
High	23.76 ± 1.89***	11.64 ± 1.01***	27.80 ± 1.10***

EE = emotional exhaustion, DP = depersonalization, PA = personal accomplishment

*** p-value is < 0.001.

Table 11.5 Correlations Between MBI Subscales and Personality Traits

	Person's rho	Extr	N	EE	DP	PA
Extraversion	Correlation Coefficient	1	–0.224	–0.297	–0.262	0.369
	Sig. (2-tailed)		0.001	0.000	0.003	0.000
Neuroticism	Correlation Coefficient		1	0.596	0.439	–0.310
	Sig. (2-tailed)			0.000	0.000	0.000
EE	Correlation Coefficient			1	0.615	–0.485
	Sig. (2-tailed)				0.000	0.000
DP	Correlation Coefficient				1	–0.351
	Sig. (2-tailed)					0.000
PA	Correlation Coefficient					1
	Sig. (2-tailed)					

Extr = Extraversion, N = Neuroticism, EE = emotional exhaustion, DP = depersonalization, PA = personal accomplishment

employees working in prisons has not been studied. Our test results confirm that the correctional officers who took part in this survey are generally burned-out. One-third of the correctional officers experience considerable emotional exhaustion, approximately more than half treat prisoners in an impersonal manner (depersonalization), and about 90% evaluate themselves negatively (reduced personal accomplishment).

The emotional exhaustion component represents the basic individual stress dimensions of burn out. The correctional officer is exposed to shift schedules that disrupt the normal sleep pattern and social life (Ranta & Sud, 2008). Working with inmates is described by a significant number of correctional staff as demanding and stressful (Finn, 1998). When correctional officers feel ineffective and powerless with inmates, they are more likely to experience emotional exhaustion. This is probably due to the never-ending demands and needs of inmates, and also the fact that some inmates are highly manipulative and sometimes oppose any help or direction from staff members; therefore, the amount of time spent each day interacting with inmates is hypothesized to be positively associated with burn out (Morgan et al., 2002). The correctional officers are also exposed to authoritarian management style having poor personal relationships with supervisor. Further lack of adequate planning and resources, lack of autonomy in performing duties and lack of recognition for work accomplishment and excessive paper work are enough to make them emotionally exhausted (Ranta & Sud, 2008). This is the feeling of being "at the end of the rope" – a feeling of being depleted of the psychological resources necessary to deal with daily tasks (Kerley, 2005).

Officers who experience burn out also tend to develop a tendency towards depersonalization. This means that they become increasingly cynical, negativistic, and even confrontational towards inmates, citizens and their fellow officers (Kerley, 2005). It develops in response to the over-load or emotional exhaustion and is self-protective at first – an emotional buffer of detached concern. But the risk is that the detachment can turn into dehumanization (Ranta & Sud, 2008). Lee and Ashforth (1990) argue that depersonalization constitutes one form of defensive behavior defined as reactive and protective actions intended to avoid an unwanted demand or reducing a perceived threat. Thus depersonalization is predicted to be associated with psychological strain and with escape as a method of coping (Huby et al., 2002).

Reduced personal accomplishment refers to a decline in feelings of competence and productivity at work. In our study more than 90% of correctional staff evaluates themselves negatively. Maslach et al. (2001) point out that the work situation with chronic overwhelming demands that contribute to exhaustion and cynicism is likely to erode one's sense of effectiveness. The lowered sense of self-efficiency is linked to inability to cope with demands of the job and it can be exacerbated by lack of opportunity to develop professionally. Persons with reduced personal accomplishment are unhappy and dissatisfied workers (Ranta & Sud, 2008).

Gender is a biographical factor that has generally been associated with burn out. Results are mixed and at times contradictory. Some studies have indicated that women are more likely to report high levels of burn out, whereas others have found the opposite (Lynch, 2007). An analysis of our data from the sample has revealed gender differences for the MBI subscales. Women show greater burn out on two aspects – higher emotional exhaustion and lower personal accomplishment, while men show greater burn out on one – higher depersonalization. Women are more likely to feel emotionally drained and overextended by their work, and less likely to feel that they have accomplished a lot in that work. Men are more likely to have a negative and callous view of people. The first argument is that women are more likely to get emotionally involved with the problems of inmates and thus to overextend themselves emotionally. A second argument is that women are more likely to be in positions involving direct contact with people such as "front-line" staff. Next, women are more likely than men to be responsible for the emotional needs of their family and thus are faced with a double dose of the strain of caring for others – both at home and on the job (Maslach & Jackson, 1985). Our results agree with the findings of some studies, indicating that men experience higher scores on depersonalization than women (Greenglass et al.,

1998; Ogus et al., 1990). One explanation is in accepted norms associated with the masculine gender role, which emphasizes strength, independence, separation and invulnerability. Another explanation derives from the emphasis on achievement, which is an integral part of the masculine gender role. If men are also competitive and their feelings of masculinity depend on successful achievement, their cynicism may derive from distrust of those with whom they are competing (Schaufeli & Greenglass, 2001). Correctional work for females remains a relatively new phenomenon in comparison to male employment standards, and it is possible that women have learned to adjust to the demands and stressors of the correctional environment. More up-to-date studies have observed no significant relationship between gender and job burn out (Griffin, 2006). To reinforce this point Carlson et al. (2003) have found that female prison officers demonstrate a greater sense of job-related personal achievement and accomplishment than their male counterparts (Regan, 2009). Future studies should also assess the differences between how males and females interact in the correctional environment.

Work-related burn out has been suggested to be a consequence of a long-term mismatch between the person's abilities or expectations and the job's characteristics (Maslach & Leiter, 1997). Some researchers expect that work environment factors are more important than personal characteristics in shaping the attitudes of correctional employees (Paoline et al, 2006). According to Maslach, the reasons for a burn out often depend not on a person but on a situation. Some personal peculiarities influence the development of this situation. Personality determines the way the person interprets events and conceives the interaction between self and the environment. One component of personality is temperament. Temperament is that aspect of our personalities that is genetically based, inborn, there from birth or even before. The role of temperament as antecedence for burn out process is supported in many studies (Cieslak et al., 2008; Kim et al., 2009; Raycheva et al., 2012; Storm & Rothmann, 2003a). For example, Raycheva et al. (2012) identify the main effects of temperament and personality traits on burn out according to the Stoyanov-Cloninger model after psychological climate dimensions are taken into account. They measure person characteristics with Temperament and Character Inventory – R, validated in Bulgarian language (Tilov et al., 2012). The findings of their study indicate that the source burn out syndome may come as much from within individuals, based on their temperamental and character traits, as from outside of them, deriving from the psychological climate dimensions.

In our study the application of the Eysenck Personality Inventory focuses on prison personnel, has shown that introversion is a significant predictor for all three components of burn out. Our results agree closely with the findings of similar studies in the field (Grundy, 2000; Jonker, 2004). Introversion is significantly correlated with emotional exhaustion (De Vries & Van Heck, 2002; Schaufeli & Enzmann, 1998). According to Storm and Rothmann (2003a) introversion seems to have a moderating effect on emotional exhaustion when job demands exceed coping resources. It makes sense that a positive correlation between emotional exhaustion and introversion has been found, because introverts have a preference for intemalising own energy and do not speak their minds as extraverts (Watson & Clark, 1997). Not ventilating own experiences of stressors, increases the risk of exhaustion. It can thus be argued that introverted correctional officers in this sample are more likely to suffer from exhaustion. A positive correlation has been found between depersonalization and introversion in our study. Correctional officials who are sociable, assertive, bold and self-sufficient are less likely to become cynical. The members of prison staff who are constantly cynical will be prone to feelings of anxiety, anger and depression. We have found that introversion is a significant predictor for reduced personal accomplishment, but the other findings are controversial. Schaufeli and Enzmann (1998) report that extraversion is related to reduced personal accomplishment, but Storm and Rothmann (2003a) find that extraversion is associated with higher personal accomplishment. A possible explanation can be that those correctional officials who are sociable, bold and self-sufficient will experience more feelings of professional efficacy. The extroverted members of staff view certain aspects of their jobs as laborious i.e. paperwork and mutine tasks. Such time consuming and laborious tasks are a far cry from the preferred activities of extraverts that include boldness, sociability, gregariousness and self-diciency. Watson and Clark (1997) note that extraverts are less dysphoric, less introspective and less self-preoccupied than introverts. According to Storm and Rothmann (2003a) introversion seems to have a moderating effect on emotional exhaustion and low openness to experience and a moderating effect on low personal accomplishment when job demands exceed coping resources. Research on the personality correlates of burn out has indicated that neuroticism was associated with (Maslach et al., 2001) and predicted burn out (Burke & Greenglass, 1995, 1996). Individuals high in neuroticism tend to express more negative emotions, emotional instability and stress reaction, and therefore they become more vulnerable to burn out and to increased psychopathology (Watson et al., 1994). High levels of neuroticism were predictive of emotional exhaustion. Storm and

Rothmann (2003a) are of the opinion that emotional stability is associated with higher personal accomplishment. Emotional stability is also related to lower depersonalization. In the case of depersonalization, neuroticism was the most important predictor, whereas personal accomplishment was predicted by low levels of neuroticism, and high levels of extraversion. Schaufeli and Enzmann (1998), in a comprehensive review of more than 250 studies on burn out, reported that neuroticism was one of the strongest personality correlates of burn out, particularly of emotional exhaustion. Personality characteristics of employees can moderate the effect of stressful situations on burn out.

11.6 Conclusion

Burn Out is a problem among prison staff. We have found high levels of three components of burn out among correctional officers – emotional exhaustion, depersonalization and reduced personal accomplishment.

Among demographic variables, only gender is significantly associated with burn out.

The personal characteristics introversion and neuroticism are the strongest personality correlates of burn out among correctional officers.

It should be noted that the results presented in this article are from only one study. There is a need for far more research on prison staff. Future studies need to assess and explore the effectiveness of burn out reduction strategies. We suggest that burn out can be prevented and reduced through the implementation of prophylactic programs in the workplace.

11.7 Acknowledgements

The current paper has been conducted with the financial support of the Scientific Research Fund of the Plovdiv Medical University under research project HO-22/2012. The authors would like to thank the correctional officers that participated in this study and the administration of the prisons for their cooperation.

References

[1] Armstrong, G. S. & Griffin, M. L. (2004). Does the job matter? Comparing correlates of stress among treatment and correctional staff in prisons. Journal of Criminal Justice 32, 577–592.

[2] Burke, R. J., & Greenglass, E. (1995). A longitudinal study of psychological burn out in teachers. Human Relations 48, 187–202.

[3] Burke, R. J., & Greenglass, E. (1996). Work stress, social support, psychological burn out and emotional and physical well-being among teachers. Psychology Health and Medicine 1, 193–205.

[4] Bühler, K. E. & Land T. (1990). Burn Out and personality in extreme nursing: an empirical study. Schweizer Archiv fur Neurologie und Psychiatrie 155, 35–42.

[5] Carlson, J. R., Anson, R. H. & Thomas, G. (2003) Correctional officer burn out and stress: Does gender matter? The Prison Journal 83, (3) 277–288.

[6] Cieslak, R., Korczynska, J., Strelau, J. & Kaczmarek, M. (2008). Burn Out predictors among prison officers: The moderating effect of temperamental endurance. Personality and Individual Differences 45, 666–672.

[7] De Vries, J. & Van Heck, G. L. (2002). Fatigue: relationships with basic personality and temperament dimensions. Personality and Individual Differences 33, (13) 11–1324.

[8] Eysenck, H. J. & Eysenck, S. B. G. (1975). Manual of the Eysenck Personality Questionnaire. London: Hodder and Stoughton.

[9] Finn, P. (1998). Correctional officer stress: A cause for concern and additional help. Federal Probation 62, 65–74.

[10] Garner, R. (2008). Police stress: Effects of criticism management training on health. [Electronic Version]. Applied Psychology in Criminal Justice 4, (2) 243–259.

[11] Greenglass, E., Burke, R. J. & Konarski, R. (1998) Components of burn out, resources and gender-related differences. Journal of Applied Social Psychology, 28, 1088–1106.

[12] Griffin, M. L. (2006). Gender and Stress: A comparative assessment of sources of stress among correctional officers. Journal of Contemporary Justice 22, (1) 4–25.

[13] Grundy, S. E. (2000). Perceived work-related stressors, personality and degree of burn out in firefighters. Dissertation Abstracts International: Section B: The Sciences and Engineering 61, 1685.

[14] Huby, G., Gerry, M., Mckinstry, B., Porter, M., Shaw, J., & Wrate, R. (2002). Morale among general practitioners; qualitative study exploring relations between partnership arrangements, personal style, and work load. British Medical Journal 325, 140–144.

[15] Jonker, B. E. (2004). Burn Out, job stress and personality traits in the South African police service. [mini-dissertation]. North-West University, Potchefstroom campus.

[16] Keinan, G. & Maslach-Pines, A. (2007). Stress and burn out among prison personnel: Sources, outcomes, and intervention strategies. Criminal Justice and Behavior 34, 380–398.

[17] Kerley, K. R. (2005). The Costs of Protecting and Serving: Exploring the Consequences of Police Officer Stress. Policing and Stress, 73–86.

[18] Kim, H. J., Shin, K. H. & Swanger, N. (2009). Burn Out and engagement: A comparative analysis using the Big Five personality dimensions. International Journal of Hospitality Management 28, 96–104.

[19] Lambert, E. G., Cluse-Tolar, T., & Hogan, N. L. (2007). This job is killing me: The impact of characteristics on correctional staff job stress. Applied Psychology in Criminal Justice 3, (2) 117–142.

[20] Lee, R. T. & Ashforth, B. E. (1990). On the meaning of Maslach's three dimensions of burn out. Journal of Applied Psychology 75, 743–747.

[21] Lynch, J. (2007). Burn Out and engagement in probationary police officers: A scoping paper. Australasian Centre for Policing Research Scope 1.

[22] Maslach, C. & Jackson, S. E. (1985). The Role of Sex and Family Variables in Burn Out. Sex Roles 12, (7/8) 837–851.

[23] Maslach, C. & Jackson, S. E. (1986). MBI: Maslach Burn Out Inventory. Manual Research Edition. Palo Alto, CA: Consulting Psychologists Press.

[24] Maslach, C. & Leiter, M. P. (1997). The truth about burn out. San Francisco: Jossey Bass.

[25] Maslach, C., Schaufeli, W. B. & Leiter, M. P. (2001). Burn Out. Annual Review of Psychology 52, 397–422.

[26] Morgan, R., Van Haveren, R. & Pearson, C. (2002). Correctional officer burn out: Further analysis. Criminal Justice and Behavior 29, 144–160.

[27] Ogus, D., Greenglass, E. & Burke, R. J. (1990) Gender role differences, work stress and depersonalization. Journal of Social behavior and Personality 5, 387–398.

[28] Paoline, E. A., Lambert, E. G. & Hogan N. L. (2006). A Calm and Happy Keeper of the Keys The Impact of ACA Views, Relations With Coworkers, and Policy Views on the Job Stress and Job Satisfaction of Correctional Staff. The Prison Journal 86, (2) 182–205.

[29] Ranta, R. S. & Sud, A. (2008). Management of Stress and Burn Out of Police Personnel. Journal of the Indian Academy of Applied Psychology 34, (1) 29–39.

[30] Raycheva, R., Asenova, R., Kazakov, D., Yordanov, S., Tarnovska, T. & Stoyanov D. (2012). The vulnerability to bur out in health care personnel according to Stoyanov-Cloninger's model: evidence from a pilot study. International Journal of Person Centered Medicine 2, (3) 552–563.

[31] Regan, S. (2009). Occupational Stress amongst Prison Officers. The Journal of Social Criminology 1, (1) 23–54.

[32] Schaufeli, W. B. & Enzmann, D. (1998). The burn out companion to study and practice: A critical analysis. London: Taylor & Francis.

[33] Schaufeli, W. B. & Greenglass, E. R. (2001). Introduction to special issue on burn out and health. Psychology and Health 16, 501–510.

[34] Storm, K. & Rothmann, S. (2003a). Burn Out in the South Afncan Police Service. Poster session presented at the 11th European Congress on Work and Organizational Psychology, Lisbon, Portugal.

[35] Tilov, B., Dimitrova, D., Stoykova, M., Tornjova, B., Foreva, G. & Stoyanov, D. (2012). Cross-cultural validation of the revised temperament and character inventory in the Bulgarian language. International Journal of Person Centered Medicine doi:10.1111/j.1365–2753.2012.01895. x/.

[36] Toch, H. (2002). Stress in policing. Washington, StateD. C.: American Psychological Association.

[37] Watson, D. & Clark, L. A (1997). Extraversion and its positive emotional core. In Hogan, R., Johnson, J. & Briggs S. (Eds.), Handbook of personality psychology (pp. 767–793) San Diego, StateCA: Academic Press.

[38] Watson, D., Clark, L. A., & Harkness, A. R. (1994). Structures of personality and their relevance to psychopathology. Journal of Abnormal Psychology 108, 18–31.

12

Coping Strategies and Burn Out Syndrome Prevention

Tanya Turnovska MD, PhD, DMSc, Full Professor,
Department of Hygiene and Ecomedicine,
Faculty of Public Health, Medical University of Plovdiv

Rositsa Dimova MD, PhD, Head Assistant Professor of Social Medicine,
Department of Health management,
Health Economics and General Practice,
Faculty of Public Health, Medical University of Plovdiv

Boris Tilov PhD, Assistant Professor of General Psychology
Department of Health Care Management,
Faculty of Public Health, Medical University of Plovdiv

Dessislava Bakova PhD, Senior Assistant Professor of Health Care
Management Department of Health Care Management,
Faculty of Public Health, Medical University of Plovdiv

Stanislava Harizanova MD, PhD Assistant Professor,
Department of Hygiene and Ecomedicine,
Faculy of Public Health, Medical University of Plovdiv

According to the World Health Organization (WHO) mental health and diseases caused by stress are among the major reasons for early retirement and premature death in Europe, (Dollard, 2003), leading to huge losses to national economies. The consequences of work-related stress include increased morbidity and impaired mental well-being of employees, associated with individual performance and workplace efficiency deterioration, absenteeism due to sick leave, decreased motivation, intentions to leave and losses to organisations as a whole, (Tennant, 2001). The development of *burn out* syndrome among employees increases organisations' direct and

Drozdstoy St. Stoyanov (Ed.), New Model of Burn Out Syndrome: Towards Early Diagnosis and Prevention, 161–170.

indirect expenses. Direct expenses include losses caused by vacancies, lost productivity, additional expenses for personnel recruitment and training due to increased manpower fluctuation, etc. Indirect expenses are formed by losses due to manpower instability, decreased productivity, stress increase and risk of *burn out* development among employees, as well as decreased responsibility of employed persons towards organisations, Bährer-Kohler (2013). Economic losses due to poor occupational health and decreased working capacity of employees may reach 10–20% of country gross domestic product. All these facts highlight the priority significance and the necessity to develop coping strategies and specific prevention of *burn out* syndrome.

There are numerous studies on the factors associated with the reduction of environmental negative impact on personality in the context of their problem interaction (Weiner, I. 2003). One of the most effective means to reduce stress levels during adaptation process is social environment support (Cohen & Wills, 1985; Wilcox, 1981). Social support which differs in terms of its type and scope has been described: organisational, horizontal, emotional, socially integrative, instrumental, etc. According to the effectiveness of its impact on individuals, researchers divide social integration into two types – having direct and buffer effect (Folkman & Lazarus, 1980; Barrera, 1988). The correlation between social support and professional burn out levels is reverse. This means that the more support individuals feel from their family and colleagues, the more able they will be to reduce stress levels effectively and quickly, (Curtona & Russell, 1990).

The study of stress-producing professional factors, together with their minimisation, avoidance or elimination from working environment are important prerequisites for effective stress prevention in the risk contingent tested. This requires the development of effective programs and strategies for the conduct of adequate events aiming at professional stress overcoming in the most vulnerable groups, consisting of medical personnel, who are subject of the study, described in other chapters of the book. In this regard, several leading medical specialists pointed out different levels of stress management in working environment. According to R. Lazarus the main coping characteristics are: to solve any problems occurred, to manage and cope with negative emotions. Two major strategies to reduce stress influence on personality are formed on this basis:

1. Problem-focused strategies, directed towards particular problem solving.
2. Emotion-focused strategies, directed towards emotional regulation.

Lazarus and Folkman created a methodology to investigate various coping strategies. The researchers established eight major factors: seek of social support, disassociation, situation avoidance, self-control, confrontation, acceptance of responsibilities, positive reassessment, smooth problem solving (Isaeva, 2008; Lazarus & Folkman, 1984; Lazarus, 1999).

Dicheva E. (2013) pointed out that prevention that is realised by means of psychological hygiene, prevention and rehabilitation has its logic, essence, consistency. In most general terms prevention includes:

- training in social (communication) skills;
- training and mastery of habits related to self-management and self-possession;
- assimilation of constructive coping technologies.

On its part, rehabilitation work with people suffering from *burn out* symptoms supposes:

- recovery of psychoenergetic potential;
- update of personal resources;
- rediscovery of professional activity meaning;
- recovery of self-confidence.

Gundersen L. (2001), Hansen N et al. (2009) established a correlation between coping methods duration (*short-term and long-term techniques*), applied with physicians and nurses, working at emergency departments and the degree of the *burn out* syndrome, which they have experienced. Some of the *short-term coping techniques*, that have been mentioned by the authors, include: crying, dreaming, consumption of food and various food additives; longer sleep in comparison to the usual one; ability to perceive work situations with a sense of humour.

A great number of studies, conducted in the CIS countries presented methods for evaluation, treatment and prevention of *burn out* syndrome (Boyko, 1996, 1999; Vorobyev & Vorobyeva, 2011). Koshlev A. (2008), expressing his own opinion and the opinion of other authors evaluated the *burn out* syndrome as a social defect, which became the 'plague of XXI century', that left even obesity at the second place. The author sees it as the final phase of 'Professional Dysfunctions' in which he included 'White Collar Syndrome' and 'Workaholism'. According to Koshlev A (2008), one of the major factors that make the situation unstable and stressful for associates in a given professional community are 'The lack of clearly defined criteria for professional activity effectiveness evaluation' and 'The inability to monitor work activity openly and objectively'. Therefore, he believes that the key point in 'Professional Dysfunctions' prevention, including the *burn out* syndrome is the development of such a system of organisation and personnel management, which would eliminate or minimise the prerequisites for its development. And the prerequisites for *burn out* syndrome development follow the huge variety of profession-specific stressogenic conditions having a social nature. This means that there is no universal prevention program for all professions that are exposed to risk. In our opinion, the correct approach is to use universal coping strategies to develop particular preventive programs depending on the particular, specific work process and occupational environment features. Potter C. (2006) had the reason to believe that the *burn out* syndrome-related problem may be solved effectively in favour of healthcare institution personnel and patients by applying a model to prevent the syndrome on *department or team level*. Moreover – the frame interventional program to prevent stress and burn out syndrome in nurses, that has been developed by Tsenova B. (2002, 2003), pointed out two approaches: basic and specific with several levels of the measures applied: individual, interpersonal, group and organisational. The increase of medical personnel authority resource through their education and self-awareness in a positive and high self-assessment will allow them to cope more easily with the aims of daily work and the challenges, posed by the healthcare system. The ambiguity in terms of healthcare professional roles results in satisfaction deterioration and stronger stress impact. (Tzenova, 2002). Similar conclusions are made by Carol, according to whom the low levels of control and autonomy determined higher burn out levels in the tested groups of specialists working at emergency departments.

An important measure for overcoming stress at work is the improved control on the working conditions, which will inevitably lead to improved security, order and favourable working environment. In agreement with the above-mentioned information, we may add that control has important impact on the other two medical care elements, as well - **the medical process** (part of which is the interpersonal relationship among personnel, as well as among personnel and patients) and **the medical result** (medical care effects, satisfaction, economic benefit), which may reduce stress to a great extent.

Another tactics which may contribute to the modification and minimisation of stress at work is the improved *knowledge and application of cognitive and behavioural models and good practices*.

The major coping strategies to prevent and restrict *burn out* syndrome development are divided into two groups: personality-centred and organisation-oriented as each particular program may also use a combination of both options (formation of the so-called 'organisational health model').

Personality-centred strategies:

- Helping employees develop their own skills to effectively overcome stress factors;
- Development of attitudes and values relating the satisfaction of life not only to the profession;
- Development of a positive self-assessment and behaviour of self-approval by means of autosuggestion;
- Establishment of balance between rest and work load;
- Taking care of health by adherence to an appropriate regimen of sleep and nutrition, relaxing exercises, physical activity stimulation;
- Gradual reduction of the number of habits that harm health;
- Stimulation of social activity conduct;
- Orientation to a hobby;
- Participation in educational trainings.

Organisation-oriented strategies:

- Application of adequate criteria for personnel selection in those professions where there is a risk of *burn out* syndrome occurrence and development.
- Clear specification of employees' responsibilities and development of realistic profession-related expectations;
- Creation of personality-supporting occupational environment;
- Stimulation of team work and improvement of communication among employees and their managers by means of appropriate and systematic feedback.
- Increased participation of employees in decision-making.
- Motivation development by provision of educational programs for qualification upgrade and career development stimulation;
- Creation and maintenance of 'healthy working environment' by means of effective work time organisation, duration, schedule, breaks, sanitary and hygiene conditions improvement, provision of personal protective equipment and holiday homes;
- Stimulation of conscientious employees by means of praises, prizes in kind, extra payment;
- Regular psychological monitoring and provision of professional psychological care to alleviate pressure;

- Organisation of group trainings;
- Monitoring and registration of sick leaves, accidents, personnel voluntary leave so that due support is rendered to the persons that need it.

In our opinion well-organised and institutionalised prevention programs of obligatory or advisable nature are much more effective in comparison to the single, individual and episodic measures for coping with negative psycho-emotional phenomena.

Moreover, it is important to pay attention to the gradual *burn out* development, which allows the due diagnosing and appropriate treatment of first symptoms and early phases to serve as a solid basis for the prevention of negative psychic and somatic consequences. In this regard *burn out* prevention begins with stress prevention. We may not always eliminate problems or prevent stress-producing events from occurring. It is important to be aware that we may change our approach towards the relationship with them, so as to rely on a successful change in our manner of reacting. Change depends not only on our desire, but also on a number of objective factors: age, sex, education, life experience, personal characteristics, expectations, health status. Cause/ causes establishment is always stated as the first step in stress overcoming. The next step includes activities to minimise stressogenic conditions and their levels of impact. When stress is directly related to social factors of the working environment, the actions that must be taken to change the situation are not easy - change of workplace, colleagues, head, etc. Therefore, the application of various strategies for a more creative approach to stress perception and management is pointed out as the third and final step in stress overcoming. Moreover, Koshlev A. (2008) posed the following rhetorical question: 'How can we benefir from *Professional burn out*?' At first glance it really seems impossible. But the author offered the managers acceptable ways out of the difficult situation:

1. Business restructuring.
2. Shift of associates from one place to another.
3. Development of new competences, etc.

In his opinion the most important thing is to exercise constant control on personnel 'burn out'. Otherwise, in case of uncontrolled development, *Professional dysfunctions* (including *burn out*) may burn the organisation out so that all fire-fighters will be helpless'. In all cases it is important to adhere to the 'happy medium', i.e. to use only those options which may not do any harm to the personnel, even a potential one.

Employees' personal attitude towards their own health is also extremely important – from the overall awareness of the problem, including terminal conditions, to the Assurance of healthy lifestyle and the application of coping strategies, including:

- healthy nutrition (in stress situations organism consumes nutrients more quickly than usual, which may lead to immune system deterioration – a condition requiring increased consumption of fruits, vegetables, whole-grain foods, etc. and/or special food additives), decrease of the consumption of animal proteins, refined foods of white flour and sugar, caffeine, alcohol, etc;
- optimisation of physical activity regimen (the everyday 30-minute energetic physical exercises reduce distress, because they 'burn out' the excessive quantities of stress hormones in blood and increase the level of endorphins – chemical substances that tend to improve mood);
- use of appropriate massage and aromatherapy with essential oils;
- herbal treatment - consumption of relaxing herbal teas (camomile, valerian, lime blossom or clover blossom) once or twice a day.
- acupressure;
- reflexotherapy – massage of reflexogenic zones of hands and feet;
- meditation;
- help by a psycho-analyst;
- others such as acupuncture, tai chi, massage and yoga, flower essences, etc.

Group practices are most appropriate for syndrome progressive development treatment and prevention, respectively, in case of already formed *burn out* syndrome phase or particular symptoms. The so-called *balint* groups, named after their founder Michael Balint (English psychoanalyst of Hungarian origin) are widely spread. Balint M. (1997) stated that the main purpose of his book is, as follows: 'to describe certain processes in the physician-patient relationship, that cause unnessecary suffering, irritation and useless efforts on the part of patients and physicians'. The main purpose of Balint seminars is to make physicians aware that they are not alone in their negative and frequently frustrating experiences with patients. The physicians in these groups talk much more easily about their feelings, about themselves, about the elimination of the problem zone in the physician-patient relationship. Balint groups generally consist of 8–12 people, they gather regularly (most commonly once in 2 weeks) and the sessions continue for 1.5 – 2 hours. Each session usually provides sufficient time for two of the physicians to present a summarised case, in which they had difficulties or in which various questions arised. The group participants receive help so that they may understand the subconscious conflicts in physician-patient relationship. Each participant must try to explain why they have reacted as they did in the respective moment. The leader must clarify the emotional situation, related to the presented case within the group process. By visiting such a seminar specialists learn to understand themselves and to control their relationship with patients. Surveys and interviews with participants in Balint seminars from all over the world showed that group discussions play a significant role in professional burn out prevention and overcoming. In our opinion it is important for medical students to study Balint approach, as well, in order to clarify their own emotional reactions before they become physicians. In most of the countries in Europe and America the participation of healthcare specialists in Balint groups is obligatorily and it is part of their student preparation and postgraduate qualification.

References

[1] Boyko V. Energy of emotions: look at myself and the others. Moscow, Sciences, 1996.

[2] Boyko V. Syndrome of "emotional burn out" in the professional converse. Sanct Peterburg, 1999.

[3] Balint M. The Doctor, His Patient and The Illness. SP Fund, NCCHI, Sofia, 1997.

[4] Vorobyov V., Vorobyova E. Negative Stress: identifying and prophylactic. Professional and organizational stress: diagnostic, prophylactic and correction. Matters of Russian national conference with international participations, town of Astrahan, 7–8 October 2011, Institute of pedagogy, psychology and social work. ISBN 978-5-9926-0509-9.

[5] Koshlev A. Syndrome "white collar" or Prophylactic of "professional burn out", 2008. ISBN 978-5-476-00603-9.

[6] Tzenova B. Major approaches to cope with burn out syndrome and stress in nurses. Health Management. 2002;1:23–4.

[7] Tzenova B. Specific approaches to cope with burn out syndrome and stress in nurses. Health Management. 2003, 2, 33–5.

[8] Bährer-Kohler, S. (2013). Burn Out for Experts: Prevention in the Context of Living and Working. Springer Science and Business Media, StateNew York.

[9] Barrera, M. J. Models of social support and life stress. In Cohen, L. H. (Ed.) Life events and psychological functioning. Newbury Park: Sage, 1988.

[10] Cohen, S., Wills, T. A. Stress, social support, and the buffering hypothesis. Psychological Bulletin, 1985, 98, 310–357

[11] Curtona, C. E., Russell, D. W. Type of social support and specific stress: Toward a theory of optimal matching. In B. R. Sarason , I. G. Sarason, and G. R. Pierce (Eds.), Social Support: An Interactional View, 1990, pp. 319–361. New York:Wiley.

[12] Dollard, M. (2003). Introduction: context, theories and intervention. In: Dollard MF, Winefield AH, Winefield HR (Eds.), Occupational Stress in the Service Professions. London, StateNew York: Taylor &Francis, 1–42.

[13] Dicheva E. The prevention of burn out syndrome in the social worker's activities. Management and education, 2013, vol. IX (4),152–157.

[14] Folkman, S., Lazarus, R. An Analysis of coping in a middle aged community sample. // Journal of Health and Social Behavior, 1980, 21, pp. 219–239.

[15] Lazarus, R., Folkman, S. Stress, appraisal and coping. New York: Springer Publishing Company Inc. 1984.

[16] Lazarus, R. S., Folkman, S. Manual for the Ways of Coping Questionnaire. Palo Alto, CA, 1988

[17] Lazarus, R. Stress and Emotion. A new Syntesis. NY , SpringerPublishing Company Inc. 1999.

[18] Potter C. To what extent do nurses and physicians working within the emergency department experience burn out: a review of the literature. Australasian Emergency Nursing Journal 2006, 9, 57–64.

[19] Tennant, C. (2001). Work-related stress and depressive disorders. Journal of Psychosomatic Research, 51: 697–704.

[20] Weiner, I. (Ed. in Chief) Handbook of Psychology., V.12, Industrial and Organizational psychology. Weiner, I. & Sons, Inc, N. Jersey, 2003.

[21] Wilcox, B. L. Social support, life stress, and psychological adjustment: A test of the buffering hypothesis. American Journal of Community Psychology, 1981, 9, pp. 371–386.

Index

About the Editors

Drozdstoj (Drossi) Stoyanov was born on July, 20th 1980 in Sofia, Bulgaria. He graduated from the high school in 1996 and received his MD from the Medical University of Sofia in 2002. He presented a PhD thesis in the field of theory and methodology of neuroscience in 2005; certified in December 2007 by the Government Specialty Board with the rank of Psychiatrist and awarded Postgraduate Certificate in Philosophy of Mental Health from the University of Central Lancashire, United Kingdom in October 2010.

Dr Stoyanov was tenured as Associate Professor in the Medical University of Plovdiv in 2008 and hold the position of Vice Dean for International Affairs of its Faculty of Public Health from 2009 to 2011; since 2011 appointed in the Faculty of Medicine, Department of Psychiatry and Medical Psychology and Special Advisor Strategic Partnerships to the Vice Rector.

In December 2013 he has been promoted to the academic position of Full Professor of Psychiatry, Medical Psychology and Person Centered Medicine.

Besides he acts as Co-Director of University Center for Philosophy and Mental Health and works as practicing psychiatrist in the 'St. Ivan Rilski' State Psychiatric Hospital.

Prof. Stoyanov was invited at the discretion of the Chair into the Philosophy Special Interest Group of the Royal College of Psychiatrists in 2007; appointed Vice Chair and member of its Executive Committee in 2012; Chair of Conceptual group in the Global Network for Diagnosis and Classification launched by the World Psychiatric Association (2008). He was elected Visiting Fellow in the Center for Philosophy of Science, University of Pittsburgh, USA in 2009.

He published more than 150 scholarly papers, including five monographs and three textbooks.

Since 2010 Prof. Stoyanov acts as Deputy Editor-in-chief of the International Annual for History and Philosophy of Medicine. He is also Associate Editor, European Journal of Person Centered Health Care (2013); Review Editor, Frontiers in Psychiatry (2013); Editorial Board member of International Journal for Person Centered Medicine (2011), Dialogues in Philosophy, Mental and Neurosciences, Folia Medica and others.

About the Authors

Assenova, Radost, MD, PhD, Associate Professor of General Practice Department of Health Management, Health Economics and General Practice, Faculty of Public Health, MUP

Bakova, Dessislava, PhD, Senior Assistant Professor of Health Care Management, Department of Health Care Management, Faculty of Public Health, Medical University of Plovdiv

Dimitrova, Donka, PhD, Engineer, Associate Professor of Health Care Management, Statistical Advisor, Department of Health management, health economics and general practice, Faculty of public health, MUP

Dimova, Rositsa, MD, PhD, Head Assistant Professor of Social Medicine Department of Health management, Health Economics and General Practice, Faculty of Public Health, Medical University of Plovdiv

Foreva, Gergana, MD, PhD, Assistant Professor of General Practice Department of Health Management, Health Economics and General Practice, Faculty of Public Health, MUP

Georgieva, Christina, MA in Clinical Psychology, PhD Candidate, Coordinator at Agency for Social Support

Haralampiev, Kaloyan, PhD, Associate Professor, Department of Sociology, Faculty of Philosophy, University of Sofia "St. Clement of Ohrid"

Harizanova, Stanislava, MD, PhD, Assistant Professor, Department of Hygiene and Ecomedicine, Faculy of Public Health, Medical University of Plovdiv

Mateva, Nonka, PhD, Associate Professor of Social Medicine, Mathematician and Statistical Advisor Department of Health Management, Health Economics and General Practice, Faculty of Public Health, Medical University of Plovdiv

Raycheva, Ralitsa, Assistant Professor, Department of Social Medicine and Public Health, Faculty of Public Health, MUP

Rozsa, Sandor, PhD, Assistant Professor, Eotvos Lorand University, Psychological Institute, Department of Personality and Health Psychology;Washington University School of Medicine, St. Louis, Department of Psychiatry, psychologist

Semerdzhieva, Mariya, MD, PhD, Associate Professor of Medical Management and Medical Etiquette, Department of Health Care Management, Faculty of Public Health, Medical University of Plovdiv

Stoykova, Maria, DD, PhD, associate professor, Head of the Department of social medicine and public health, Department of social medicine and public health, Faculty of public health, Medical University of Plovdir

Tilov, Boris, PhD, Assistant Professor of General Psychology Department of Health Care Management, Faculty of Public Health, Medical University of Plovdiv

Tornyova, Bianka, PhD, Assistant Professor of General Psychology Department of Health Care Management, Faculty of Public Health, MUP

Toshkova-Hristozova, Slavka, PhD, Department of Language and Specialized Training, MUP

Turnovska, Tanya, MD, PhD, DSc, Full Professor, Department of Hygiene and Ecomedicine, Faculty of Public Health, Medical University of Plovdiv

Lightning Source UK Ltd.
Milton Keynes UK
UKOW02n0843151214

243137UK00001B/50/P

9 788793 102705